ONLY ONE
EARTH

Books by Barbara Ward

The West at Bay
Policy for the West
Faith and Freedom
The Interplay of East and West
Five Ideas That Change the World
India and the West
The Rich Nations and the Poor Nations
Nationalism and Ideology
Spaceship Earth
The Lopsided World
Only One Earth (with René Dubos)
The Home of Man
Progress for a Small Planet
Who Speaks for Earth (with others)

ONLY ONE
EARTH

THE CARE AND MAINTENANCE OF A SMALL

PLANET BY *BARBARA WARD* AND *RENÉ DUBOS*

An Unofficial Report Commissioned by the
Secretary-General of the United Nations Conference on the Human
Environment, Prepared with the Assistance of a 152-Member
Committee of Corresponding Consultants in 58 Countries

W · W · NORTON & COMPANY
New York · London

First published as a Norton paperback 1983

Library of Congress Cataloging in Publication Data
Jackson, Barbara (Ward) Lady, 1914–1981
Only one earth.
"An unofficial report commissioned by the Secretary-
General of the United Nations Conference on the Human
Environment, prepared with the assistance of a 152-
member committee of corresponding consultants in 58
countries."
1. Human ecology.   I. Dubos, René Jules, 1901–
joint author.   II. United Nations Conference on the
Human Environment, Stockholm, 1972.   III. Title.
GF41.J3   301.31   72-447
ISBN 0-393-30129-X

W. W. Norton & Company, Inc., 500 Fifth Avenue, New York, N.Y. 10110
4 5 6 7 8 9 0

CONTENTS

Part Four: The Developing Regions

Part Five: A Planetary Order

PREFACE

THIS REPORT is the result of a unique experiment in international collaboration. A large committee of scientific and intellectual leaders from fifty-eight countries served as consultants in preparing the report, of whom more than seventy made detailed written contributions directly to the work of preparing it.

The names of Barbara Ward and René Dubos are listed quite properly as authors of the report. They are indeed responsible for the drafting and revision of the manuscript to which they both contributed at personal sacrifice, under remorseless time pressure, with unstinting assistance of a very small staff, and without compensation. They are responsible, too, for its general style. It would be quite impossible to describe adequately the spirit and energy that they invested in its enterprise.

But in this case the role of the "authors" is more accurately described as creative managers of a cooperative process—one which engaged many of the world's leading authorities as consultants in the multiple branches of environmental affairs. Their names appear elsewhere.

As Secretary-General of the United Nations Conference on the Human Environment, I commissioned Dr. Dubos in May 1971 to serve as chairman of a distinguished group of experts who would serve as advisers in preparing the report. The aim was to reach out for the best advice available from the world's intellectual leaders in providing a conceptual framework for participants in the United Nations Conference and the general public as well. Members of the group of consultants were asked to read a preliminary manuscript and offer their criticisms and contributions. The letter appointing Dr. Dubos stated that the greatest value of the report would "derive precisely from the fact that it will represent the knowledge and opinion of the world's leading experts and thinkers about the relationships between man and his natural habitat at a time when human activity is having profound effects upon the environment."

This report has been considered an integral part of preparations for the United Nations Conference. At the same time it is the work of individuals, serving in their personal capacities without restraints imposed upon officials of governments and international agencies. Thus the report is not an official United Nations document but a report *to* the United Nations Conference Secretariat from an independent expert group. The only restraint on those who prepared the report was a request that they not prejudge the work of governments at the United Nations Conference by proposing specific international agreements or actions—its main purpose being to provide background information relevant to official policy decisions.

Many were skeptical of the workability of the procedure adopted for preparation of this report. Yet with less than thirty days to study the preliminary draft, prepare their comments, and return them to New York, more than seventy such contributions were received in time to be considered in the course of revising the manuscript. Almost without exception, the comments from the group of experts were substantive, specific, and constructive. Many were lengthy and detailed.

As the authors stress in the introduction, there are contrasting views about the social application of important categories of available technology even where the scientific facts are not in serious dispute. In other cases viewpoints expressed by consultants in effect canceled each other out by recommending that the drafters give countervailing weights to various factors and considerations. Some found the tone too alarmist; others found it too optimistic. All this is very valuable because it is just as important for the decision-maker to know that experts disagree as it is to find that a consensus exists. It also means, inevitably, that everyone cannot be satisfied at the same time; perhaps none of the contributors will be *wholly* pleased with the final text, certainly not the few whose useful comments unfortunately arrived too late to be taken into account. But I know that the managers of this difficult creative process have made every effort, under the most pressing circumstances, to find a balance among frequently contrasting views.

More specifically, the consultants provided a great deal of invaluable guidance in the formulation of scientific issues in suggesting rearrangement of material and in verifying or correcting factual points.

The United Nations Conference Secretariat is not responsible for the content of this report, nor is it called upon to endorse it in whole or in part. But the Secretariat does welcome enthusiastically the success of the collaborative process by which it was produced and expresses the deepest appreciation to members of the consulting group and to all those many people who, in one way or another, aided in this remarkable process.

Finally, I must offer profound thanks to the Albert Schweitzer Chair at Columbia University, the World Bank, and the Ford Foundation for fully funding the cost of this report. The International Institute for Environmental Affairs provided highly effective overall management of a complex process with no precedents to provide guidance.

> Maurice F. Strong
> Secretary-General
> United Nations Conference on the Human Environment

INTRODUCTION

THIS INTRODUCTION is distilled from about four hundred pages of correspondence, originating from forty different countries. It is inspired by the letters we received in reply to our request both for criticisms of the preliminary draft of *Only One Earth* and for suggestions as to what should be emphasized in the final text. From the tone of the letters, many of which greatly exceed ten pages in length, it is clear that most of our consultants are intensely worried about the state of our planet, but that very few if any of them regard the situation as hopeless. The vibrant concern of so many thoughtful and learned persons, from many different parts of the world and different fields of human endeavor, is enough reason for sober optimism.

We are, of course, immensely grateful to our consultants for calling to our attention factual errors, omissions, and misplaced emphases in the preliminary draft. But the most rewarding and illuminating aspect of their replies was the diversity and richness of the conceptual points of view they expressed concerning the problems to be discussed by the U.N. Conference on the Human Environment. The very ambiguity of the phrase "human environment" clearly provided the consultants with the opportunity to formulate their social and scientific philosophies and to explore the consequences of their attitudes in operational terms.

The spectrum of views among our consultants was much wider than we expected; but far from resulting in confusion, the diversity of their attitudes toward the environment turned out to be the expression of the richness of man's nature—and it is this richness which accounts for the diversity of civilizations. Free human beings differ not only with regard to the characteristics of the environmental settings which they find most desirable, but also with regard to life-styles, aspirations, and last but not least their views of man's place in nature. Experts as well as laymen usually find it easy to agree on purely objective scientific issues. But the U.N. Conference is not focused on abstract problems of theoretical ecology. It is primarily concerned with the characteristics of the environment which affect the quality of human life—a very subjective and ill-defined concept.

In his reply, one of the consultants from Africa urges us to spell *Man* with a capital *M,* instead of writing about man or men. In our opinion, this is not trivial stylistic advice. It symbolizes rather a conceptual problem which inevitably confronts environmentalists in all their practical discussions and decisions. Are men simply higher apes, and as such of no greater significance than other components of the natural ecosystems? Or does *Man* occupy a special place in nature?

Those of our consultants whose primary interest is theoretical ecology naturally urge that emphasis be placed on the earth ecosystem per se, man being considered chiefly as a disturbing element in it. And there is no doubt indeed that most of our present environmental difficulties originate from man's ecological misbehavior. Increasingly we consider ourselves not as lodgers on the earth, but as its landlords; we identify progress with the conquest of the external world even if this means destruction of those parts of nature which we assume—often erroneously—to be irrelevant to our welfare. But while it is possible that *Homo sapiens* could survive as a biological species after impoverishing and spoiling nature, could *Man* long retain his humanness in a desecrated environment?

The statesmen who planned the U.N. Conference on the Human Environment certainly had in mind the physical and spiritual qualities of man's relation to the earth, at least as much as the ecological health of our planet. They were naturally preoccupied with the shortages of food and amenities, the depletion of natural resources, the accumulation of environmental pollutants, the increase in the world population, and also the threat to certain natural values which transcend bodily needs. They realized in addition that all these problems have acquired an element of extreme urgency from the fact that mankind has now spread over the whole surface of the globe. By the year 1985, according to recent estimates, all land surfaces will have been occupied and utilized by man except for those areas which are so cold or at such high altitudes that they are incompatible with continued human habitation or exploitation.

The U.N. Conference on the Human Environment comes therefore at a very critical time. Now that mankind is in the process of completing the colonization of the planet, learning to manage it intelligently is an urgent imperative. Man must accept responsibility for the stewardship of the earth. The world *stewardship* implies, of course, management for the sake of someone else. Depending upon their scientific, social, philosophical, and religious attitudes, environmentalists have somewhat different views as to the nature of the party for whom they should act as stewards. But in practice the charge of the U.N. to the Conference was clearly to define what should be done to maintain the earth as a place suitable for human life not only now, but also for future generations.

The depletion of natural resources is, of course, one of the chief reasons of uncertainty concerning the continued ability of the earth to support future human civilizations. Concern about future supplies of natural resources is so widespread and so deep that one of our consultants, from a highly industrialized affluent European country, went as far as suggesting that mankind must begin very soon to retreat from industrialization and to focus efforts on the development of more efficient agricultural techniques! Thoughts of retreat from industrialization, however, are not congenial to the consultants who belong to parts of the world which are only now beginning to industrialize in order to lift themselves from poverty. They are aware of the dangers inherent in industrialization, but they see it as the only road to higher living standards. In fact, almost any method of industrial development which gives hope of more abundant food production, less unemployment, better public health, and a decent level of amenities must have in their judgment precedence over considerations of future environmental damage.

Since industrial growth depends upon the availability of large amounts of electric power and of certain chemical products, it is not surprising that policy-makers and planners from countries which are seeking economic development are not likely to be sidetracked, in the words of an Asian statesman "by dreams of landscapes innocent of chimney stacks." There is indeed a widespread acceptance of the fact that environmental pollution is an inescapable by-product of industrial development. Experience shows furthermore that societies have become preoccupied with long-range ecological consequences only after industrialization had provided them with a high level of economic affluence. "Sufficient unto the day is the evil thereof" has been the law which has so far tacitly governed much of man's behavior toward the environment. If history repeats itself in this regard, it is likely that in most places and for many years, environmental quality will be subordinated to developmental goals.

Economic affluence, however, is only one of the factors affecting civic consciousness in its attitude toward the environment. The difficulty of settling by scientific expertise the comparative importance of technological and environmental considerations in industrial development is well illustrated by the profound differences of views among our consultants regarding nuclear power.

On the very same day we received forceful statements on nuclear power from two Nobel laureates, both equally illustrious for the magnitude of their achievements in the natural sciences and for the importance of their social contributions as leaders of national agencies and as advisers to international bodies. Both, furthermore, are from highly industrialized English-speaking countries. According to one of them, the text of *Only One Earth* does not do full justice to the potentialities of nuclear power and greatly exaggerates its threats to natural ecosystems and to human health; in contrast, the other Nobel laureate affirms that nuclear power should not be developed at all, because, in his words, it is "utterly inappropriate in the biosphere." Many other consultants have expressed equally strong views on both sides of this controversy.

As could be expected, similar contrasts of opinion repeatedly occur among the consultants with regard to pesticides. One of them informs us that he probably would be dead if DDT had not been available at the time he was working in Guyana; in the same vein, many others repeatedly assert that millions of people will soon die of infectious disease or of malnutrition if attempts are made to limit drastically the use of pesticides in public health practice and in agriculture. There are many other experts, on the other hand, who are convinced that natural ecosystems are even now profoundly disturbed by pesticides and who predict that the earth will progressively become unsuitable to human life if present trends of pesticide use are continued.

A highly important but confusing anthology could thus be compiled from the spectrum of views sent by our consultants regarding the effects of technological intervention into the human environment:

· Some are more impressed by the stability and resilience of ecosystems than by their fragility.

· Some would emphasize human settlements rather than natural ecosystems and nature conservation.

· Some would give priority to water pollution, others to the state of the atmosphere, still others to the problems of land management.

· Some believe that environmental pollution and the depletion of natural resources can best be controlled by individual behavior, others by strict controls over industry, and still others by a complete transformation of the political structure or of life-styles.

· Some believe that the most destructive forms of ecological damage flow from types of high-energy, high-profit technology whose advantages are grossly overstated in terms of genuine utility, others see energy as *the* key to the basic economic achievement of producing more goods for fewer inputs and thus incomparably widening the citizen's wealth and choice.

· Some see the solution of environmental problems in more scientific knowledge and better technological fixes, others in socio-economic morality, and still others in the cultivation of spiritual values.

· Some object to the phrase "developed countries" because they believe that no part of the world is yet adequately developed; others in contrast believe that industrial development has gone too far in the affluent countries and must be reduced within limits determined by man's ability to stabilize the economy of the earth's resources. As mentioned earlier, certain consultants from highly industrialized countries go as far as advocating a return to an economy based on agriculture and believe the developing countries would be unwise to regard technology as the way to the future.

There was general agreement among the experts that environmental problems are becoming increasingly world-wide and therefore demand a global approach. But two consultants from two different Asian countries suggest that little progress will be made, either in economic development or in environmental improvement, until each particular country has learned to manage its own ecosystem. As they point out, there are many different worlds within our theoretical One World, each differing from the other not only in physical characteristics and economic structure, but even more importantly perhaps in cultural traditions and in aspirations.

Some of the consultants feel that the general tone of *Only One Earth* is far too pessimistic and they see no justification in reporting on the present state of the world as if it were a "fear story." One of them, indeed, sees in the style all the defects he violently objects to in *Silent Spring*— "emotional and non-factual." Other consultants, in contrast, would like the book to issue a more forceful warning—a clarion call—that present environmental trends cannot continue much longer because mankind is on a course of self-destruction. One consultant specifically urges the authors of *Only One Earth* not to let the editorial staff reduce the book to a mere recital of facts because salvation will ultimately depend on an emotional awakening.

The list of conflicting views and recommendations received from our consultants could be extended to many pages. It constitutes a spectrum of expert opinion on environmental improvements which ranges all the way from the advocacy of technological fixes to a plea for new religious attitudes. At first sight, this discrepancy of opinions appears to constitute evidence for the commonly held view that experts do not agree on facts and therefore are of little help in formulating programs of action. But in reality experts rarely disagree on the validity of the facts themselves; they differ only with regard to the interpretation and use of these facts.

No one doubts, for example, that ionizing radiations increase mutation rates, that most mutations are deleterious, and that some damage to human life and to ecological systems is therefore likely to result from the increase in radiation level—small as it may be—that will inevitably be caused by the operation of large numbers of nuclear power plants. But while all scientists agree on these facts, individually they differ as to the levels of radiation they consider tolerable, because this involves social considerations based on value judgments. For example, the biological hazards resulting from the industrial use of nuclear power must be balanced against the advantages to be derived from the economic development made possible by this power. Needless to say, similar arguments could be developed for most other technological innovations.

The problem of value judgment is further complicated by the fact that, in addition to the initial effects of technological interventions, there commonly occur indirect and delayed consequences which are difficult to predict and to evaluate. DDT causes little if any direct and immediate damage to man when used under reasonably controlled conditions. Its toxicity for the large ecosystems of nature, and eventually for man himself, becomes evident only after prolonged periods of usage which result in its progressive accumulation in the food chains. Technological interventions must therefore be judged not only from the point of view of their effects in the here and now, but also with regard to the possibility that they will affect man, or his environment, or both at some later date. The U.N. Conference on the Human Environment could serve the very useful purpose of highlighting the need to focus social and scientific attention on the indirect and often unpredictable delayed responses made by complex ecosystems to social and technological innovations.

Since policies concerning the human environment require both social judgment and specialized scientific knowledge, perceptive and informed laymen can often contribute as much as technical experts to their formulation. In certain cases, indeed, laymen may be wiser judges than experts because their overall view of the complexity of human and environmental problems is not distorted by the parochialism which commonly results from technical specialization.

The diversity of views held by experts, even within a given social system and a given nation, points to the nature of the difficulties that will certainly face the delegates at the U.N. Conference on the Human Environment. In most cases, the difficulties will originate not from uncertainties about scientific facts, but from differences in attitudes toward social values.

The establishment of a desirable human environment implies more than the maintenance of ecological equilibrium, the economical management of natural resources, and the control of the forces that threaten biological and mental health. Ideally it requires also that social groups and individuals be provided with the opportunity to develop ways of life and surroundings of their own choice. Man not only survives and functions in his environment, he shapes it and he is shaped by it. As a result of this constant feedback between man and environment, both acquire distinctive characteristics which develop within the laws of nature, yet transcend the blind determinism of natural phenomena. The exciting richness of the human environment results not only from the immense diversity of genetic constitution and of natural phenomena but also and perhaps even more from the endless interplay between natural forces and human will.

Ambassador Adlai Stevenson clearly had in mind the overpowering influence of man's role in determining the quality of the environment and therefore of human life when, in his last speech before the Economic and Social Council in Geneva on July 9, 1965, he referred to the earth as a little spaceship on which we travel together, "dependent on its vulnerable supplies of air and soil." We are indeed travelers bound to the earth's crust, drawing life from the air and water of its thin and fragile envelope, using and reusing its very limited supply of natural resources. Now that all habitable parts of the globe are occupied, the careful husbandry of the earth is a *sine qua non* for the survival of the human species, and for the creation of decent ways of life for all the people of the world. The fundamental task of the U.N. Conference on the Human Environment is to formulate the problems inherent in the limitations of the spaceship earth and to devise patterns of collective behavior compatible with the continued flowering of civilizations.

It is deliberately that, in the last paragraph, we have used the word *civilizations* in the plural. Just as individual human beings differ in their life and aspirations, so do social groups. This is obvious from the wide range of views—often so far apart as to seem incompatible—expressed by the consultants to the Report on the World Environment. But far from being a reason to despair, this divergence of views is in fact the expression of one of the most appealing aspects of the human species— its diversity. There are possibilities within the human environment for many different kinds of surroundings and ways of life.

While collaborating with a large international group in the preparation of *Only One Earth*, one of us (René Dubos) was simultaneously working on another book in which he emphasizes the importance of developing the distinctive genius of each place, each social group, and each person—in other words of cultivating individuality. These two endeavors are not incompatible; in fact, they correspond to two complementary attitudes. The emotional attachment to our prized diversity need not interfere with our attempts to develop the global state of mind which will generate a rational loyalty to the planet as a whole. As we enter the global phase of human evolution it becomes obvious that each man has two countries, his own and the planet Earth.

COMMITTEE OF CORRESPONDING CONSULTANTS

THE FOLLOWING INDIVIDUALS agreed to serve as consultants and were invited to offer their criticisms and comments on an early draft of *Only One Earth*. The appearance of their names here, in grateful appreciation of their constructive help, does not constitute an endorsement of the book.

Acronyms for international organizations used in identifying the consultants:

ACAST	Advisory Council on the Application of Science and Technology
CERN	European Center for Nuclear Research
CIAP	Inter-American Committee of the Alliance for Progress
ICSU	International Council of Scientific Unions
ISSC	International Social Sciences Council
IUCN	International Union for the Conservation of Nature
OECD	Organization for Economic Cooperation and Development
SCOPE	Scientific Committee on Problems of the Environment
SID	Society for International Development
UNCTAD	United Nations Conference on Trade and Development
UNESCO	United Nations Educational, Scientific and Cultural Organization

Rufus O. Adegboye, Department of Agricultural Economics and Extension, University of Ibadan; member of Council, SID. *Nigeria*

Malcolm S. Adiseshiah, Director, Madras Institute of Development Studies; former Deputy-Director General, UNESCO. *India*

Anwar Ahmed Aleem, Institute of Marine Biology and Oceanography, Fourah Bay College, University of Sierra Leone. *Egypt*

Ismail Al-Azzawi, Director, Agricultural Technical Institute, Baghdad. *Iraq*

Fouad Ammoun, International Court of Justice, The Hague. *Lebanon*

Serge Antoine, Secretary-General, High Committee on the Environment, Paris. *France*

J. A. Armstrong, President, Imperial Oil Ltd., Toronto. *Canada*

Edgard Julius Barbosa Arp, President, Grupo Industrial Arp, Rio de Janeiro. *Brazil*

Sir Eric Ashby, Master of Clare College, Cambridge; Chairman, Royal Commission on Environmental Pollution. *United Kingdom*

Pierre Victor Auger; Director, Laboratory of Space Physics; formerly Director-General, European Space Research Organization; member, ACAST. *France*

Jean G. Baer, biologist; Institute of Zoology, University of Neuchatel. *Switzerland*

Moisés Béhar, Director, Institute of Nutrition for Central America and Panama, Guatemala City. *Guatemala*

Daniel Bell, Professor of Sociology, Harvard University. *United States*

Fredrik Berglund, medical scientist, National Institute of Public Health, Stockholm. *Sweden*

Carl Gustav Bernhard, President, Royal Academy of Sciences, Stockholm. *Sweden*

Edouard Bonnefous, President, French Committee, European League for Economic Cooperation; President, Academy of Moral and Political Sciences. *France*

François Bourlière, biologist; Professor of Gerontology, University of Paris; President, Special Committee for the International Biological Program, ICSU. *France*

Harrison Brown, Foreign Secretary, National Academy of Sciences; Professor of Geochemistry, California Institute of Technology. *United States*

Silviu Brucan, Professor of Sociology, University of Bucharest; former Ambassador to the United Nations and the United States; member, Pacem in Maribus. *Romania*

M. F. Mörzer Bruyns, Professor of Applied Ecology, Nature Conservation Department, Agricultural University, Wageningen; Vice President, IUCN. *Netherlands*

Colin Douglas Buchanan, Professor of Transport, Imperial College, London; formerly Urban Planning Adviser, Ministry of Transport. *United Kingdom*

Sir Macfarlane Burnet, immunologist; Professor Emeritus, University of Melbourne; Past President, Australian Academy of Science; Nobel Laureate. *Australia*

Louis Camu, Chairman, Bank of Brussels. *Belgium*

Franco A. Casadio, Chairman, Executive Committee, World Federation of United Nations Associations, Rome. *Italy*

Carlos Chagas, Director, Institute of Biophysics, Federal University of Rio de Janeiro; member and former Chairman, ACAST. *Brazil*

Wilbert K. Chagula, Minister for Water Development and Power, Dar-es-Salaam; member, ACAST. *Tanzania*

D. A. Chant, Chairman, Department of Zoology, University of Toronto. *Canada*

Pradisth Cheosakul, Secretary-General, National Research Council, Bangkok. *Thailand*

LaMont C. Cole, Professor of Ecology, Division of Biological Sciences, Cornell University. *United States*

Barry Commoner, Director, Center for the Biology of Natural Systems, Washington University, St. Louis. *United States*

Gamani Corea, Deputy Governor, Central Bank of Ceylon, Colombo. *Ceylon*

Pierre Dansereau, Scientific Director, Research Center in World Ecology, Montreal; Vice President, International Biological Program. *Canada*

J. C. de Graft-Johnson, Lecturer in Sociology, University of Ghana; Honorary Secretary, Ghana Academy of Arts and Sciences, Accra. *Ghana*

Francesco di Castri, Director, Institute of Ecology, Austral University of Chile, Valdivia; Vice Chairman, SCOPE. *Chile*

Constantinos A. Doxiadis, architect; Doxiadis Associates International, Consultants on Development and Ekistics, Athens. *Greece*

P. Duvigneaud, Professor of Botanical Sciences, University of Brussels. *Belgium*

Rolf Edberg, journalist and diplomat; Governor, Province of Varmland, Karlstad. *Sweden*

Laila Shukry El Hamamsy, Professor of Anthropology; Director, Social Research Center, The American University, Cairo. *Egypt*

Fujio Egami, President, Science Council of Japan; Professor of Biochemistry, University of Tokyo. *Japan*

Arne Engström, Professor, Department of Medical Physics, Karolinska Institute, Stockholm. *Sweden*

Donald S. Farner, Professor of Zoophysiology; Chairman, Department of Zoology, University of Washington, Seattle; Executive Officer, ICSU; President, International Union of Biological Sciences; member, SCOPE. *United States*

Frank John Fenner, research biologist; Director, John Curtin School of Medical Research, The Australian National University, Canberra. *Australia*

Joseph L. Fisher, President, Resources for the Future, Inc., Washington, D.C. *United States*

Sir Otto Frankel, geneticist; formerly Chief, Division of Plant Industry, Commonwealth Scientific and Industrial Research Organization, Canberra. *Australia*

Sir Frank Fraser Darling, biologist; Vice President, IUCN; member, Royal Commission on Environmental Pollution. *United Kingdom*

Henning K. Friis, economist; Director, Danish National Institute of Social Research, Copenhagen; member, Executive Committee, ISSC. *Denmark*

Hans-Walter Georgii, Institute for Meteorology and Geophysics, Johann Wolfgang Goethe University, Frankfurt. *Federal Republic of Germany*

I. P. Gerasimov, Director, Institute of Geography, Soviet Academy of Sciences, Moscow; President, All Union Society of Soil Sciences. *U.S.S.R.*

Walery Goetel, Full Member, Academy of Mining, Krakow. *Poland*

Jeffrey Gustavson, Co-chairman of White House Conference on Youth and Environment. *United States*

Jermen M. Gvishiani, Deputy Chairman, State Committee for Science and Technology of the USSR Council of Ministers; member, ACAST. *U.S.S.R.*

Knut Hagrup, President, Scandinavian Airlines System, Stockholm. *Norway*

Philip Handler, biochemist; President, National Academy of Sciences. *United States*

Kurt Hansen, Chairman of the Board, Farbenfabriken Bayer; Chairman, Supervisory Board, Agfa-Gevaert; member, Supervisory Board, Siemens. *Federal Republic of Germany*

Jorge E. Hardoy, Director, Center of Urban and Regional Studies, Buenos Aires. *Argentina*

Ilmo Hela, Director, Institute of Marine Research, Helsinki; member, Executive Board, UNESCO. *Finland*

Thor Heyerdahl, anthropologist and explorer; Vice President, World Association of World Federalists. *Norway*

Jonathan Holliman, former President, International Youth Federation for Environmental Studies; Staff, 1971 International Youth Conference on the Human Environment. *United Kingdom*

W. David Hopper, agricultural economist; President, International Development Research Centre, Ottawa. *Canada*

Sidney Howe, President, The Conservation Foundation, Washington, D.C. *United States*

Enrique V. Iglesias, economist; member, Advisory Group of Experts, CIAP. *Uruguay*

Neil Iliff, Deputy Chairman, Shell Chemicals U.K. Ltd., London; member, Royal Commission on Environmental Pollution. *United Kingdom*

M. Kassas, Professor of Botany, University of Cairo, Giza. *Egypt*

Ichiro Kato, educator, lawyer; President, University of Tokyo; Chairman, National University Association. *Japan*

Anna Medwecka-Kornaś, Nature Conservation Research Center, Krakow; member, Executive Board, IUCN. *Poland*

F. A. Mehta, Director, Tata Industries Ltd., Bombay. *India*

Hermann Meusel, botanist; Martin Luther University, Halle. *Democratic Republic of Germany*

Martin Meyerson, urbanologist; President, University of Pennsylvania. *United States*

G. L. Monekosso, Director, University Center for Health Sciences, The Federal University of Cameroun, Yaoundé. *Cameroun*

Jerome Monod, specialist in land management; Administrator, National Company of the Rhone Valley. *France*

Giuseppe Montalenti, Director, Institute of Genetics, University of Rome; President, Commission for the Protection of Nature, Italian Research Council. *Italy*

Lewis Mumford, writer on urban planning; Bemis Professor Emeritus, Massachusetts Institute of Technology. *United States*

Karl Gunnar Myrdal, Professor of International Economy, Institute for International Economic Studies, University of Stockholm. *Sweden*

Shigeo Nagano, Chairman, Board of Directors, Nippon Steel Corporation, Tokyo; Vice President, Tokyo Chamber of Commerce and Industry. *Japan*

Imre V. Nagy, Professor of Hydraulic Engineering, Department for Hydrology, Agricultural Water Management, Water Supply and Sewerage, Technical University, Budapest. *Hungary*

E. M. Nicholson, Chairman, Land Use Consultants, London; Convener, The Nature Conservancy. *United Kingdom*

Letitia E. Obeng, Director, Institute of Aquatic Biology, Achimota; Co-manager, Volta Lake Project. *Ghana*

Thomas R. Odhiambo, Professor of Entomology; Director, International Centre of Insect Physiology and Ecology, Nairobi. *Kenya*

Saburo Okita, President, Japan Economic Research Council, Tokyo; member of Council, SID. *Japan*

Pitamber Pant, Planning Commission, New Delhi. *India*

Arvid Pardo, Center for the Study of Democratic Institutions, Santa Barbara; former Ambassador of Malta to the United Nations; member, Pacem in Maribus. *Malta*

Aurelio Peccei, Vice Chairman, Olivetti and Company; President, Club of Rome. *Italy*

Oliverio Phillips Michelson, industrial consultant, Bogotá; former member, ACAST. *Colombia*

Jacques Piccard, economist and oceanographer, President of Council, Foundation for the Study and Protection of the Sea and Lakes; member, Pacem in Maribus. *Switzerland*

Sarwono R. Prawirohardjo, President, Indonesian Institute of Sciences, Djakarta; former member, ACAST. *Indonesia*

Raúl Prebisch, Director-General, Latin American Institute of Economic and Social Planning, Santiago; former Secretary-General, UNCTAD. *Argentina*

Dixy Lee Ray, Director, Pacific Science Center Foundation, Seattle. *United States*

Robert Reichardt, Professor of Social Philosophy, University of Vienna; member, Executive Committee, ISSC. *Austria*

Roger Revelle, Director, Harvard Center for Population Studies, Cambridge; member, Pacem in Maribus. *United States*

Benito F. Reyes, President, University of the City of Manila; Chairman, World Congress of University Presidents. *Philippines*

Lord Ritchie-Calder of Balmashannar, journalist and science writer; Professor of International Relations, University of Edinburgh. *United Kingdom*

Victor L. Urquidi, economist; President, College of Mexico; member, ACAST. *Mexico*

Fred van der Vegte, President, International Youth Federation for Environmental Studies; member, Planning Committee, 1971 International Youth Conference on the Human Environment. *Netherlands*

Claudio Veliz, Director, Institute of International Studies, University of Chile, Santiago. *Chile*

Alexander P. Vinogradov, Director, Vernadsky Institute of Geochemistry and Analytical Chemistry; Vice President, Academy of Sciences. *U.S.S.R.*

C. F. Freiherr von Weizsacker, Director, Max Planck Institute, Starnberg; former Professor of Philosophy, University of Hamburg. *Federal Republic of Germany*

Joseph S. Weiner, educator; Convener, Royal Anthropological Institute, London; Professor of Environmental Physiology, University of London; member, SCOPE. *United Kingdom*

Carroll Wilson, Professor, Alfred P. Sloan School of Management, Massachusetts Institute of Technology; Director, Study of Critical Environmental Problems. *United States*

Mohamed Yeganeh, Alternate Executive Director, International Bank for Reconstruction and Development, Washington, D.C.; member, ACAST. *Iran*

Lord Solly Zuckerman, anatomist; Honorary Secretary, The Zoological Society of London; former Chief Scientific Adviser to the U.K. Government. *United Kingdom*

ONLY ONE
EARTH

Part One: The Planet's Unity

1 THE WORLD WE INHERIT

Man Makes Himself

MAN INHABITS TWO WORLDS. One is the natural world of plants and animals, of soils and airs and waters which preceded him by billions of years and of which he is a part. The other is the world of social institutions and artifacts he builds for himself, using his tools and engines, his science and his dreams to fashion an environment obedient to human purpose and direction.

The search for a better-managed human society is as old as man himself. It is rooted in the nature of human experience. Men believe they can be happy. They experience comfort, security, joyful participation, mental vigor, intellectual discovery, poetic insights, peace of soul, bodily rest. They seek to embody them in their human environment.

But the actual life of most of mankind has been cramped with back-breaking labor, exposed to deadly or debilitating disease, prey to wars and famines, haunted by the loss of children, filled with fear and the ignorance that breeds more fear. At the end, for everyone, stands dreaded, unknown death. To long for joy, support, and comfort, to react violently against fear and anguish is quite simply the human condition.

To some extent, these reactions can be found in other animals. Birds weaving nests, beavers building dams, animals hunting in packs are altering, "improving," and safeguarding their lives and their environments in a purposeful way. Man shares with his animal forebears many of the responses required for dealing successfully with a natural world that is at once beneficial and destructive. The original brain was an efficient receiver of sensation and director of appropriate emotional and sensory responses in the rest of the body—running from fire, cowering from attacking beasts, embracing and love-making.

It is with the final stage in the brain's development that man, as man, begins to draw away from his ancestors. At some point, probably about a hundred thousand years ago, the forebrain became enormously larger and more complex. The skull of modern man is three times larger than the so-called Australopithecine hominid who is generally thought to be man's immediate predecessor. This change in size and structure of man's brain increases his ability both to receive sensations and to engage in abstraction, reflection, forethought, and the rational choice of goals. To serve abstract thought alone, we are told, the brain contains ten thousand times more components than the present generation of most complex computers. And the computer is yet to be invented that also smells, tastes, sees, and touches, thus adding to its capacity for abstract thinking all the emotional richness and complexity of a total human response.

This extraordinary development of man's brain lessens his dependence on animal instinct but is at the root of his creativity and his destructiveness. He can modify, more drastically than any bird or beaver, the conditions he finds unsuitable. And, having found his first experiment unsuccessful, he has far more immediate freedom to cast about and try something new. But he can also carry his experiments to disastrous points of no return from which instinctive reactions might have held him back.

This freedom has its counterpart. Some form of order must be imposed on such a range of possibility and risk. No social unit, even one as small as the family, can live on permanent change, innovation, and experiment. The instinctive response had to be supplemented by elements of a man-made social and physical design—first for self-preservation and then for all the extra dimensions of meaning—beauty, safety, and utility—which man could now conceive and therefore, in varying degrees, realize. From the start of his existence, man has innovated—in social forms, in technical improvements. His condition is to live aspiringly and uncertainly where the biosphere of living things and the technosphere of his inventions interact.

But today, as we enter the last decades of the twentieth century, there is a growing sense that something fundamental and possibly irrevocable is happening to man's relations with both his worlds. In the last two hundred years, and with staggering acceleration in the last twenty-five, the power, extent, and depth of man's interventions in the natural order seem to presage a revolutionary new epoch in human history, perhaps the most revolutionary the mind can conceive. Men seem, on a planetary scale, to be substituting the controlled for the uncontrolled, the fabricated for the unworked, the planned for the random. And they are doing so with a speed and depth of intervention unknown in any previous age of human history.

The Beginnings of Innovation

Scale and speed are the keys to this revolution. If we look back over the pattern of man's millennial history, we can detect from man's earliest beginnings an underlying acceleration in both the variety of his interventions and the pace at which they succeed each other. This is not an order of "progress" in the optimistic sense of the eighteenth and nineteenth centuries. Good and bad are distributed all along the way. Some of the most fruitful inventions long precede other less fortunate ones. But it *is* a progression in the whole scale of man's ability to change his environment for both good and ill.

His first invention may be his greatest. It is language itself, the ability to communicate with other human beings through the symbols of speech, through words which are sounds to which agreed meanings have been attached. They made possible the organized activities of groups and clans. They underlay common strategies for the chase and the snare. They were at the beginnings of incantation and ritual, of poetry and the telling of tales. For tens of thousands of years, language has been man's most useful tool.

At some more recent date, a new and formidable intervention begins —the use of nonhuman energy to enhance human activity. In very early times, man learned to use animals to help him perform his work. But with the use of fire he begins his experiments with the earth's vast sources of nonanimal energy.

He is the first creature not to flee from fire. Perhaps he first used it in the hunt to frighten animals into the open. Perhaps one day, in hunger, he tasted a burnt animal that had not fled in time. Cooking began round the campfires, and the fire which had been tamed from raging energy in the dark forests and dry grasses would become in time the symbol of the hearth and the center of household use and comfort.

Fire, too, played its part in early man's greatest technical innovation

—the invention of settled agriculture. To this day in many subsistence economies, the basic farming technique is "slash and burn." The ash from burning trees enhances the soil. When fertility is used up and crops begin to decline, the clan moves to another part of the forest; trees grow again and their leaves eventually form new humus in the resting soil. This was one of the techniques by which some ten thousand years ago, in various parts of the planet, men learned to imitate nature's cycles of growth and thus began to free themselves from their millennial dependence upon hunting and food gathering.

It was, in fact, a period of incomparable inventiveness. Tools which, in the shape of sticks for shoving and picking or of rocks for weapons, man derives from his animal forebears were now refined by chipping or fashioning into the knives and axes and hoes of the Stone Age. Houses were built. Cloth and containers came from the newly invented loom and the potter's wheel. Brewing began. Cooking grew more varied and venturesome; the hearth warmed the house in colder climates.

Fire took men, too, beyond purely agricultural and domestic uses by making possible the metal ages. Once again, the chance sight of a surface metal melting in a charcoal fire may have first pointed the way to malleable metal for human use. Man could cease chipping and grinding stone and move to the smelting of metals. The Bronze Age and then the Iron Age followed the Stone Age. The relative durability of the new materials multiplied their uses. Every kind of implement became more sophisticated and versatile. So did decoration and embellishment. The instruments of the chase acquired new effectiveness. So did the weapons of war, the iron sword cleaving the bronze shield. And here, at a very early stage in man's use of nonhuman energy in developing technology, we encounter a strange and archetypal warning.

Fire helped to clear the forests and fertilize the fields. Fire smelted the metals. Fire warmed the hearths. Its use, helping to provide a surplus above mere subsistence, prepared the ground for the first large-scale experiments in organized civilization—in the Middle East, in North India, in China. Yet in one of the earliest Western mythologies fire is not a beneficent gift. It is stolen from the gods and Prometheus, the thief, is chained to the bare rock with a vulture at his vitals to avenge the blasphemous act. With this new power and ability to shape his environment, man is seen in Greek mythology to be playing a godlike part, creator, innovator, remaker of his world and of himself. This is his dignity and freedom. But it is potentially the way to overweening pride and to an arrogance readily spilling over into the risk of destruction.

The Early Civilizations

The scale and pace of man's interventions increased further with each development and elaboration of civilized life. Early civilizations were based with few exceptions on river valleys whose waters were managed so as to bring reliable water supplies to the farms (still a major preoccupation in river-valley management). These vast river systems—on the Nile, the Euphrates, the Indus, and the Yellow River—required complicated administration and engineering to ensure their successful working. Bureaucracies grew up, vocations became more differentiated, a written language was needed since face-to-face consultation could no longer take place over such wide domains (many of the earliest written documents are inventories of palace and temple stores). Money was developed to take trade beyond the stage of village barter. Commerce opened the land routes and sea routes between Asia and the Middle East. Cities grew up round court and temple. Bureaucrats, traders, artisans moved to the centers of power. Above all, the management of waters required reliable measurements of land and flood and exact knowledge of times and seasons. Mathematics and astronomy were born among the Chaldeans and the Egyptians and later gave birth to the Greek vision of universal law which would embrace ultimate reality.

By the time the Han dynasty had come to power in China, and Rome had begun to assert its imperial control in the Mediterranean some 2100 years ago, civilized societies commanded most of the instruments of organization and technology that were to last man through another thousand years. They had alphabets and mathematical measures. They could use fire and water and winds and tides to supplement animal energy. They had learned to use a whole range of metals. They had elaborated on all the domestic and agricultural arts of neolithic man. They had cities and bureaucracies. They had coins and trade. This was the technological heritage upon which human society was to be very largely based for more than another millennium. Napoleon's land armies went no faster than Hannibal's. Charcoal still smelted iron ore until the eighteenth century. Water wheels powered the first factories. The Arabs knew as much mathematics as Galileo.

But in the seventeenth century the tempo begins once again to quicken. For a couple of hundred years, every index of growth—population, energy, use of food supplies, minerals consumed, people leaving the countryside and clustering in the cities—begins to mount. Many of the estimates are still guesses, but for population trends, use of energy, and increase in urbanization, they are probably not too far off the mark.

The Hinge of History

Then, in the twentieth century, as the accompanying charts illustrate, every index takes off for the stratosphere. Energy use, the consumption of foodstuffs and raw materials, urbanization, above all, population—every one of them seems to spring off the end of the graphs. Here, clearly, we confront one of those increases and accelerations in which quantitive changes are so great that they constitute a qualitative change. The whole human way of life is, as it were, pulling at its anchors in nature and in

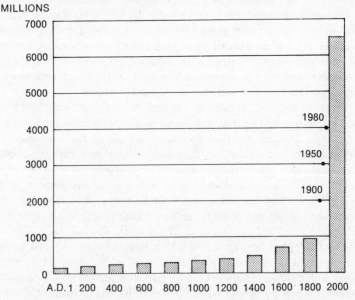

WORLD POPULATION A.D. 1—2000

Source: United Nations Data

history and straining to set sail. Or should we rather say that it is gathering energy on its launching pad to take off, rocketlike, into regions as relatively uncharted as the surface of Mars?

At the core of the new situation is the interaction of increasing numbers of people, all using or seeking to make use of more energy and more materials, all tending to draw together in closer proximity in urban regions, all concentrating to a wholly new degree the by-products of their activities—their demands and consumption, their movements and noise, their wastes and effluents. The graphs give us some idea of the dimensions. World population edged up from the levels made possible by neolithic agriculture to perhaps 400 million by the fall of Rome. Over a thousand years later, about A.D. 1600 it reached the first billion. Thereafter the acceleration begins as a result of rising production in farms and factories as the Industrial Revolution gathered momentum and was followed by a steady fall in the death rate, particularly in infant mortality. The second billion arrives after only three hundred years, in 1900. The third took only fifty years. We are well on our way to the fourth in only thirty years—by 1980.

This rate of growth of population in the twentieth century has been accompanied by the settlement of virtually all the naturally inhabitable parts of the globe and an increase of more than a billion people in urban settlements of over 20,000 inhabitants, by a quadrupling of energy consumption, and a virtually uncountable increase in the consumption of depletable resources. Today it has been estimated that, on the average, a citizen in the world's wealthiest country—the United States—carries eleven tons of steel around with him in cars and household equipment and produces each year one ton of waste of all sorts. Even these brief indications are enough to show that the impact of man and his technology on his natural environment and resources is already radically different from anything yet experienced in human history.

But this is only the beginning. If we extend our predictions only another thirty years, we have a likely world population of seven billion people. Urban inhabitants, at nearly three and a half billion, will outstrip rural people for the first time. Energy use will be thirty times greater than in 1900 and may have quadrupled since 1970. This, however, is simply an extrapolation from existing levels of use. But the two-thirds of the world's people who live in developing lands consume nearly eight times less energy per capita than do citizens in richer lands. Can we be sure

WORLD ENERGY CONSUMPTION
A.D. 1900—2000 IN MILLIONS
OF METRIC TONS OF COAL EQUIVALENT

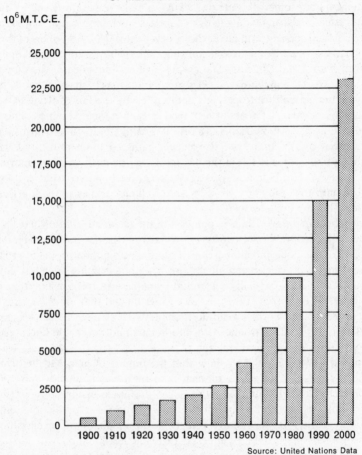

Source: United Nations Data

that their demands may not undergo as startling an expansion? Is it conceivable that the next century may begin with seven billion people commanding, say at least half the energy use, food and metal consumption, and output of effluents reached today in the United States?

Before we dismiss the idea as a fantasy, it is well to reflect for a moment upon one fundamental factor in the energy-consumption equation. It is a fairly general characteristic of human nature that men seek to avoid backbreaking and monotonous work, that they like comfort, are fascinated by personal possessions, and enjoy having a good time. The

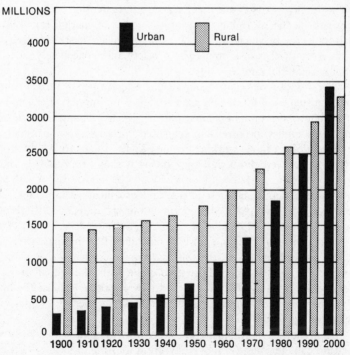

WORLD URBAN*AND RURAL POPULATION

Source: United Nations Data

*Urban refers to towns of more than 20,000 people

proof of this basic psychological bias can be seen in the behavior of any wealthy group ever since neolithic man began, through settled agriculture, to build up a surplus of goods above the level of tribal subsistence. That this bias can carry with it high costs in terms of boredom and triviality is not in doubt. But the point is that making three-quarters of the population affluent, as opposed to the traditional 1 per cent, will not make them less likely to want the things wealthy people normally want —very little menial work, a profusion of goods, and lots of chances for entertainment.

The reason why the modern age has seen, in a number of countries, an extension of wealth from a traditional elite to a much larger number of citizens is in part political. It stems from the emergence of equality as a general ideal—even if it is still fairly far from an achieved practice. But wider prosperity is due much more to the extensions of technology; above all, to the enormous increase in supplies of energy. Energy is at the root of the productivity, of the ability to make "more for less" that offers most citizens in a modernized society an inconceivably enlarged range of material choice.

One way of looking at this large expansion in personal opportunity has been suggested by Buckminster Fuller, who, thirty years ago, made an estimate of the amount of muscular energy needed to produce the then available supplies of power and suggested that each American had the equivalent of 153 slaves working for him. Today, the figure would probably be nearer 400 slaves and they do what slaves traditionally did— lighten domestic work, cook food, carry people about, rush in with fans and heaters, deliver clothes, finery, and ornaments which they have produced in the first place, play continous music (loudly or softly according to command), and remove garbage from the immediate vicinity. They are no longer men. They are powered machines. It is the space they take up, the power they use up, the wastes they give up that is at the core of some of the most pressing short-term problems in the human environment—the problems of pollution. But they are there because the mass of people want their "energy slaves" and find the experience of personal wealth an agreeable one.

We do not know whether those who now enjoy these standards will want ever higher ones—a move, say, from four hundred to a thousand energy slaves over the next twenty years—although the past behavior of rich groups does not suggest that appetite grows less with eating. We

cannot be certain whether societies which have modernized their economies by the route of public ownership and central planning will undergo
quite the same pressures for increasing personal ease and consumption
as do market economies—although socialist governments undoubtedly
include a steadily rising standard of living among their national objec-
tives. Similarly, we cannot be absolutely certain that the modernizing
"South" of our planet will pursue so strenuously the goal of personal
wealth—although in many societies the attitude of the elite hardly suggests a total rejection of the high-consumption model.

What *is* certain is that our sudden, vast accelerations—in numbers,
in the use of energy and new materials, in urbanization, in consumptive
ideals, in consequent pollution—have set technological man on a course
which could alter dangerously and perhaps irreversibly, the natural systems of his planet upon which his biological survival depends. Today
when only a third of humanity has entered the technological age, the
pressures are already apparent. Rivers have caught fire and burned their
bridges. Lakes and inland seas—the Baltic, the Mediterranean—are under threat from untreated wastes, many of which can feed bacteria and
algae; these in turn exhaust the water's oxygen and threaten other forms
of marine life. The burning of fossil fuels is increasing, with unforeseeable
consequences for the earth's climates and atmosphere. Dust and particles
in the atmosphere may also alter the earth's temperature in unpredictable ways. Even the vast oceans, covering 70 per cent of the globe and
providing an apparently inexhaustible reserve of moisture, an endless
dump for wastes, and a perpetual source of freshening winds and currents, are far more vulnerable to man's polluting activities than had been
assumed. Run off into them too many poisons, insecticides, and fertilizers, void too many oil bilges, choke too many of the estuarine waters
where the fish spawn and multiply, and even the oceans may cease to
serve man's purposes as effortlessly and reliably as he now seems to
suppose.

And all these risks are appearing on the human horizon with a world
population of less than four billion, at least half of whom have hardly
raised their claims on the planet above those of neolithic man. But
suppose seven billion try to live like Europeans or Japanese? Suppose
they seek American standards of automobile use and add the emission
of three and a half billion cars to the carbon monoxide in the air and the
lungs? Suppose three-quarters of them move to the cities, seeking there

the developed world's levels of energy use and materials consumption? There is no way in which such equations can be worked out. But in that case, what "gives" on the collision course? Numbers? Yes—but whose? Consumption? Yes—but where? Urban amenities? Yes—but in which lands? Energy slaves? Yes—but not mine. Or does the planet itself, with its precious, irreplaceable, and finite resources of air and water and soil, come under increasing and even irreversible pressure?

In short, the two worlds of man—the biosphere of his inheritance, the technosphere of his creation—are out of balance, indeed potentially in deep conflict. And man is in the middle. This is the hinge of history at which we stand, the door of the future opening on to a crisis more sudden, more global, more inescapable, and more bewildering than any ever encountered by the human species and one which will take decisive shape within the life span of children who are already born.

2 AN UNINTENDED ORDER

The New Knowledge

WHAT BROUGHT mankind to this threshold? In one sense, the forces at work are as old as human society—the search for usable knowledge, the need for production and exchange, the organizing power of the social community. Early clans experimented with pits and snares, captured and shared the game, and followed the leadership of the hunting chief. These three factors—knowledge, economics, political power—became more interrelated, powerful, and self-reinforcing with each advance in technology and organization. What happened after the sixteenth century is that an unprecedented change in the scale of human activity and in the interactions between knowledge, economics, and power produced results which no one foresaw and no one intended.

Take first the question of knowledge. By the seventeenth century an immense corpus of exact and useful knowledge, based upon millennia of observation and practice, already existed in human society. In India and China, in particular, the technical skills needed for the production of textiles, pottery, porcelain, and metal goods were so far ahead of Europe's that the first aim of European merchants since the fourteenth century had been to break the monopoly of Arab merchants and middlemen and trade directly with the fabled Indies and far Cathay.

The sudden surge in knowledge in Western Europe after the sixteenth century cannot therefore be explained by an already established technological pre-eminence—although medieval innovations had been considerable. It was above all a change of approach and method. This development had three fundamental elements. The first was a wholly new emphasis on *useful* knowledge—what Francis Bacon called knowledge "for the benefit and use of man, for the relief of man's estate." Since usefulness implies reliability, the second element was the idea of the repeatable controlled experiment. In this lay the germ of science's master procedure—the movement from hypothesis, an imaginative construction which suggests a solution, to the solidity of proof, which is the hypothesis checked by every test the questing scientist can devise—but still open to further modification or disproof if new facts come up to challenge it. The third element was to devise means of reliability.

René Descartes invented one—the reduction of fields of study to their ultimate components, the "discrete objects" which make them up, irrespective of all the variables of changing situation and context. The other was the exact observation of how objects behaved so that they could be made to behave more usefully. This turned out to imply, above all, measurement. All changes imply the intervention of some force or energy. Things left to themselves continue in the same state. This inertia had long since been remarked by the philosophers. How, then, did material objects change and how could they be made to change in ways determined by man? The idea of utility implied the purposive use of energy. In fact, energy or force can only be defined as "the capacity to produce work." And force, as a capacity to pull, push, lift, drop, and generally move things about, can be precisely measured. The key to controlled use was therefore measurement. As Leonardo da Vinci had prophesied, mathematics would become the key to mechanics and indeed to all the natural sciences.

Energy and Measurement

In the seventeenth century Isaac Newton discovered and measured the gravitational force of the sun and planets in the solar system. Fahrenheit's thermometer gave man the first means of exactly measuring that most ancient of his servants—heat—and opened the way to the precise degree of "cooking" to which materials had to be exposed to produce any chemical alteration and hence any later development of a chemical industry. By measuring steam and putting it to work more exactly, James Watt powered the cotton looms in Lancashire in the eighteenth century and, as it were, formally inaugurated the Industrial Revolution. Sixty years later, steam was adapted to work in moving engines by George Stephenson.

In the nineteenth century, the pace of measurement and successful experiments quickens sharply. There is more exact knowledge from which to begin. The scientists can make larger and more assured raids into unconquered territory, consolidating their work as they advance. We can in particular follow two paths of research which prepared the way for the most Promethean of all human discoveries in the twentieth century—nuclear power.

The first path starts with Descartes's clue of looking for nature's "discrete objects" or ultimate elements. Between 1770 and 1870, a Frenchman, Antoine Lavoisier, an Englishman, John Dalton, and a Russian, Dmitri Mendeleev, managed to discover the constant and relative weights of the whole table of elements—from hydrogen* the first and lightest, to uranium, the ninety-second and heaviest. (Eleven heavier man-made elements have since been added.)

The other path lay through the clarification of the long-noticed fact of electrical energy—the phenomenon of an electric charge passing between positive and negative poles in a gravitational field of force. In the nineteenth century James Clerk Maxwell postulated the passage throughout the whole of space of these charges, "undulations" or pulses. Heinrich Hertz first observed one and Guglielmo Marconi took out the first patent for transmitting a "radio wave" and launched the world on the development of universal instantaneous communication.

But these discoveries did more than bring distant lands and then other planets into audible and—later—visible contact with the earth's inhabitants. The electric pulses could be directed down wires made of good conductive metals like copper, and a massive source of clean, usable, and distributable energy, electricity, came into being. But more than this, when the full range of all the waves in the solar system was finally put together, it was discovered that a vast unified organization of energy existed in the cosmos—the electromagnetic spectrum stretching from electrical household current (with oscillations in the neighborhood of 60 cycles per second), radio and television waves, up through the sounds of speech, the visible spectrum of observable colors, and on to cosmic rays of

*To give some idea of the infinitesimal scale of these ultimate components, one should recall that two hydrogen atoms compose one hydrogen molecule. One ounce equals 8.54×10^{24} (or 8,540,000,000,000,000,000,000,000) molecules. In mathematical shorthand, the small 24 to the right of the ten registers the number of places one moves the decimal point to the right.

such inconceivable energy as to produce 10^{22} cycles per second. From these two master discoveries—the table of elements and the electromagnetic spectrum—there followed in the twentieth century the ultimate breakthrough to the understanding of nuclear structure, the nature of atomic energy, and the beginnings of its use for human purposes in peace and war.

In studying fundamental substances, scientists discovered a number of them—fluorine, then pitchblende, then uranium—which gave off x-rays or radioactive pulses which at 10^{17} to 10^{20} cycles per second were clearly in the upper "solar" end of the spectrum and shared some of the searing power of the sun's own energy. This evidence of near-cosmic power throbbing in the core of supposedly indivisible elements was the clue to the fact that they were not—as had been supposed hitherto—minute, irreducible objects like marbles or billiard balls but immensely complicated electrical phenomena in which electrons orbited continuously round a central nucleus. Moreover, it was discovered that if they were bombarded by other atoms they would disintegrate or unite, giving off immense amounts of energy in the process. From this conclusion it was possible to infer that related processes might be at work at the heart of the sun, releasing the inconceivable floods of energy which powers all moving and living things in the entire solar system. Nuclear power, first exploded on earth at Los Alamos in 1944, brought mankind to the brink of a new meaning to the Promethean myth. Now his fire had in truth been stolen from the sun god and the theft carried with it more than a hint of the primeval curse. Misused for unchecked power and greed, it could bind the modern Prometheus to the lifeless rock in an irradiated planet.

The vast range of scientific achievements that has flowed from the precise measurements of energy and the study of closely delineated "discrete objects" makes up one of the most remarkable odysseys of the human mind. At each new stage of measurement, new kinds of energy were put at man's disposal—steam, then electricity, finally nuclear power. And with each new extension, man's power to manipulate and change "discrete objects" for the relief of man's estate was correspondingly increased. It is difficult to overestimate the degree to which the use of energy and the manipulation of materials has reduced the crushing burdens of physical work, lessened the concentration of human effort on food production, freed men for other pursuits, and extended to millions a wealth and opportunity formerly enjoyed by the smallest elite.

But until relatively recently, the underlying thrust of science's ultra-powerful operations was toward dissection and ever-increasing specialization. New scientific approaches which move within wider fields of relevant relationships—astrophysics, biochemistry, social anthropology—have not yet a full century of practice behind them. The work of synthesis has begun but has not yet offset the risk that emphasizing exact and precisely delineated spheres of study may make too little of the unities and systems underlying, connecting, and in part explaining natural events.

And this fact is closely linked to the dual nature of the scientific enterprise. On the one hand, it is dedicated to the highest standards of objectivity and enslaved to the sternest mistress—the spirit of truth. But once the results are known—the horsepower measured, the atom bombarded, the nucleus split—vastly increased powers of use and misuse fall into men's hands. Energy and matter have, as it were, been torn out of the restraints imposed upon them by the natural system. The molecular structure of material things abstracted for individual use is no longer acted upon by a myriad subtle yet systematic influences of the total environment. Stripped down to its components, it can be made more usable—as in plastic—or infinitely more destructive—as in nerve gas. Nuclear fusion in the sun is shielded from planet Earth by all the intervening layers of atmosphere and cloud cover. On earth, the defenses must be provided by men.

It follows that the missing checks and balances, the lost restraints of the system as a whole, have now to be replaced by human insight and human wisdom. But the truth is that man's tremendous scientific advance over the last three hundred years has been accompanied not by increased insight and wisdom but by equally powerful, uncoordinated, and thrusting developments in economics and politics in the pursuit of goods and the claims of sovereignty. It has not been restraint or reflection that has chiefly presided over the emergence of the new scientific, technological order. To go back to Francis Bacon, the new gods have been "the idols of the market and the idols of the tribe."

From Commerce to Industry

The new technological order is solidly rooted in man's desire for goods and his willingness to work and plan and invest to get them. It is therefore not surprising that a world-wide expansion of commerce preceded by at least a century the first large-scale technological transformations of agriculture and industry. And very probably the transformations could not have taken place without the stimulus of the new international market.

When one thinks of the extraordinary commercial predominance of the maritime nations of Western Europe in the eighteenth and nineteenth centuries, there is irony in the fact that their earlier drive to trade with the Orient stemmed from a large sense of inferiority—their lack of precious metals and stones, their backwardness in silk-making and porcelain, above all, the total absence of spices to make salted and sometimes stinking meat palatable by the end of the winter.

But the long voyages from Europe to the Spice Islands, to India, and China did more than satisfy Europe's eager contemporary consumer demands. The journeys were distant and risky. They had to be financed long before the homecoming ships, escaping scurvy, pirates, and hidden reefs, could reach port and make handsome profits on Ming China or nutmeg. Loans, credit, interest rates, risk capital, insurance, banking procedures, partnerships, and profit sharing—these prerequisites of the later development of large-scale industrial production were either invented or vastly enlarged by maritime world trade.

When, in the eighteenth century, scientific experimenters of the stamp of James Watt began to put new measurements, tools, and engines into the workshops of traditional ironmasters and weavers, they straightforwardly took over from mercantile practice the techniques of the "long journey"—no longer across distant seas, but across the extended time needed for building a larger factory and assembling a bigger work force. Capital was raised, partners brought in; their contribution paid for all the costs in the construction of the enterprise—the masons, the toolmakers, the bricks and mortar, the interest on borrowed funds. It paid the costs of getting the raw materials and the laborers to the site. It paid the first bills for wages. Then, hopefully, a consumer good—cloth, pans, kettles, pottery—would be ready for the market where consumers were waiting to buy. The price they were willing to pay for the goods had to cover all the previous costs. But the very great increase in productivity —more output for each man-hour of work—made possible mainly by adding steam power to the machines, meant that compared with artisan handicrafts the system began to pour out an enormously increased stream of regular and reliable supplies of consumer goods at reduced costs—yards of cloth, hammers, pins, cups. The goods could be sold more cheaply, more people would buy them, the costs would be recovered and a profit remain at the end which ambitious spinners or bleachers or cutlers could reinvest in more machines, more output, and more sales. On this balance between increasing supply at lower costs on the one hand, and a widening demand for basic goods on the other—with profit as reward, inducement, and margin for reinvestment—the whole of the early industrial system was built up.

As a decentralized way of satisfying a million different tastes and needs, the market system could hardly be matched. If consumers wanted less, or prices were too high, demand would fall and producers would move out of that line. Eager demand, on the contrary, would send up prices, draw in more producers, and, as a result, send up the output of those goods. Capital would flow to industries which, by satisfying demand more economically, could make the largest margin over cost. By responding promptly to market signals, the whole system would produce, through profit, an increasing surplus for reinvestment. Production would grow, markets widen further, and an expanding balance between supply and demand would bring more and more people within reach of at least a modest affluence. The whole system seemed to have an inner consistency and spontaneous effectiveness which alone could respond to diverse, fluctuating, unpredictable consumer demands and mobilize equally diverse and often uncertain productive resources.

The Price of Prosperity

Yet this immensely powerful expansion of the market, which has steadily gathered momentum for the last two centuries, has also created unintended, fragmented, and destabilizing side effects. The most obvious is the immediate impact on income distribution. If everything were to have a price, expressing a new concept of cost, a whole range of social arrangements designed for a less cost-conscious social order would disappear. The servants, who stood about for prestige's sake in the manor or the merchant's house, were dismissed. Improving farmers—in Britain in the eighteenth century, in Europe in the nineteenth, in Asia in the twentieth century's "Green Revolution"—cleared peasants and cottagers off the land and began to use new productive techniques in order to push up output per acre while cutting labor costs. The safety net of charity and mutual obligation in the village community broke. The dispossessed moved to the new industrial communities, there to compete wages down to mere subsistence and leave the entire surplus to the wealthy—long established or newly enriched—for reinvestment or for lavish consumption. The social imbalance, the gap between rich and poor, actually widened in the early stages of industrialism—as it still can do in developing societies today.

Moreover, by placing overwhelming emphasis on the sales of goods and the profit to be made from them, the system proved to be deficient in the funds and institutions needed to supply the essential *public* goods —health, education, decent city design, public safety, environmental improvement—which did not offer benefits that could be easily broken down into salable goods and in any case were too expensive for the poorer purchasers. The possibility of what was later to be called "private affluence and public squalor" was present from the beginning in economies so heavily oriented toward the production and exchange of consumer goods.

We can see the effects of this imbalance most clearly in two areas— in the early establishment of what were to come to be accepted as normal industrial costs and in the pattern of the first industrial cities. First, on the side of costs, as the Industrial Revolution gathered momentum two centuries ago, the new managers and owners adopted a concept which was not so much a conscious policy as a response to the risks they foresaw in their untried enterprise. These risks were real enough. A useless invention, a chemical that burned instead of bleached, faulty engines which belied the promise of productivity by running out of steam —these were uncertainties enough without adding extra obligations of environment or welfare. Above all, each experiment—into a new line or a new method—could lead to bankruptcy for oneself, one's family, and one's friends. There was no limited liability in the early days, and nineteenth-century novels are full of dramas of failed banks, absconding clerks, and rascally partners.

To the entrepreneur, therefore, it seemed folly to permit any avoidable costs to be laid to his account. The definition of *costs* in the early industrial system took on a *minimum* content which it has to some extent retained. Costs were what the entrepreneur could not avoid paying. Anything else was left to others or left undone. Untended slag heaps piled up beside the mines and furnaces—some of the size of the tragic mass at Aberfan in Wales, which more than a century later slid down into the valley and obliterated the village school with the children in it. Industrial effluents were left to flow off in the rivers. The factories belched choking smoke into the air above, and their interiors were as unadorned as economy could make them, nightmares of noise, heat, and danger from throbbing, thrusting, uncaged machines into which pauper workers, often under ten years old, could fall to their death.

Political interventions, such as factory acts and inspectorates, little by little improved internal conditions. But the use of air and waterways as giant sewers for effluents aroused less concern, in part because the scale of industrialization and the flow of consumer goods were not at first so great as to make winds and waters and tides incapable of clearing up a good deal of the mess. Natural systems were treated as "free goods" because they still appeared to be costless cleansers. It is true that a few of the classical economists discussed the problem of what they called "external diseconomies"—one factory's soot dirtying the next factory's windows or upstream chlorine poisoning downstream fish—in other words, the costs an enterprise can cause to others while escaping damage itself. The economists also suggested remedies—fines for delinquent behavior, taxes on effluents. But as late as 1967, one of the world's most widely read economic textbooks discussed such external diseconomies only in footnotes and an appendix.

The insufficiency of the market approach is equally evident in the early growth of the industrial cities. Technology and the market relentlessly speeded up the processes of urbanization. For the first time in human history, the bulk of man's work began to be done not on the land but in built-up areas. It was in the resulting spread of industrial plant and urban settlements that some of the most evident drawbacks of economic expansion guided exclusively by market signals were—and in some cases still are—to be found.

Actually, a tendency to centralize power and order in the new nation-states had begun to blow up the size of the capital cities ahead of technological revolution and market growth. By the sixteenth century, London had 250,000 inhabitants, Milan over 200,000, Antwerp and Amsterdam over 100,000 each. Two centuries later London had nearly a million, Paris over 650,000. But it was this expansion in the cities that began to demonstrate to the new entrepreneurs the value to business of large concentrations of people. A sizable labor force, an accessible market, speedy suppliers—all these cost-reducing elements went into the urban pattern. Industry began to move to the fringes of the largest cities. It built up its own new industrial towns, pushing Manchester from a village of some 12,000 inhabitants in 1717 to a rapidly expanding town of 30,000 only forty years later to a big industrial city of over 300,000 inhabitants by the 1860s.

In these new agglomerations the evils of wastes and effluents were enormously multiplied by concentration. And these were further increased by the poverty and desperate overcrowding of the urban workers. In these first phases of industrialization, labor costs, like all other costs, were held to a minimum. Dispossessed cottagers, handicraft workers put out of business by the new machines, the paupers, the orphans—their competition for jobs saw to it that wages did not do much more than keep alive their malnourished bodies. This poverty in turn determined much of the pattern of housing in the industrial cities. Rents had to match the meager wages. Yet in contemporary philosophy, house-building would be done only if it showed a profit. The solution was crowding. The mean little houses were built back to back in square blocks which permitted the maximum packing in of room space, half of which had access to light only through the door into the front room. In some quarters there were not even wells to provide water. Stinking closets in the basement—or none at all—failed to answer people's sanitary needs. In the streets filth accumulated.

Yet the sheer number of rent-paying families, sometimes one family per room—or per cellar—made profits for landlords and greatly inflated the capital value of urban land, thereby increasing still further the difficulty of expanding the supply of cheap houses. Casual lodgings in which workers slept in the beds in turn—as they still do in Bombay—railway arches, park benches filled in the gap.

In such environments, sickness and death were the visible companions of daily life. Death rates in the cities remained consistently higher than in the countryside and kept a check on the rate of population growth. Choking smoke and pea-soup fog proved lethal to people with respiratory diseases. Typhus and other vermin-spread diseases took a terrible toll in dirty, crowded quarters. The dreaded typhoid spread from the steaming slums to the outer world of middle-class comfort and even into the palaces of royalty.

All these evils of the urban condition set off another decisive trend in the living habits of modern man. Anyone who could get away from the dirt, disease, and noise of the inner industrial city began to do so. By the middle of the nineteenth century, the move to suburbia was under way. First the villas in their leafy gardens were within reach by horse and carriage—for instance, such early suburbs as Clapham and Hampstead in London, St. Cloud in Paris, or Brooklyn Heights in New York. Then railways allowed for greater distances and the "spread city" composed of people escaping from the central urban core began to move out in waves to ever-further settlements. Thus they tended to re-create the conditions they had tried to avoid. By the end of the century Brooklyn Heights was a city quarter like all the other "first-generation" belts of suburbia. But the spread went on, land values rising to tempt the private landseller and the questing developer and, in some lands, removing the affluent, able to pay rates and taxes, from any role in solving the problems of the older city quarters. The suburban flight thus underlined the condition that Benjamin Disraeli defined in nineteenth-century Britain as the coexistence of "two nations—the nation of the rich and the nation of the poor."

The Consumer Revolution

The evils of early industrialism and the attendant squalor of its urban order were too gross to be accepted without protest and violence. It took more than a century of constant political pressure, ever-renewed efforts of social reform, and some major political revolutions to discredit the idea that an infallible market, simply by responding to a vast variety of consumer signals, could achieve acceptable patterns of income distribution or an adequate level of general welfare. By the second half of the twentieth century, the shape of a new "social contract" had begun to emerge. The role the market should play was either subordinated under socialism to state planning or, in mixed market economies, balanced by a wider acceptance of the community's political responsibility for a whole range of public goods and the citizen's basic right to income, education, work, and welfare.

But, by a paradox, it is precisely this achievement that has created new pressures on the economic system and reinforced the difficulty of a number of earlier problems. A twenty-five-year boom set in among the industrialized nations after 1945. Highly developed consumer goods, often by-products of war-developed technologies—in energy, in electronics, in transport, in synthetics—added enormously to the range of power-using machines available to the citizen, particularly the automobile. Consumer demand was stimulated in many countries by a new intensity of advertising, the income behind it sustained by the political commitment to economic growth and full employment. This surge of demand began gobbling up resources and increasing requirements of materials and energy at an unprecedented rate. The economy responded rapidly to these signals. New energy-producing equipment, was installed, materials mobilized, sold, used, and junked. World trade doubled each decade. In many countries, the growth of national income was not far behind.

But behind the feverish expansion, old pressures began to gather momentum and new ones were added. In already developed countries, it is true, only certain social groups were desperately poor but their plight was made all the more intolerable by the contrast with the affluence of everyone else. In the world at large, the industrialized powers increased their lead in production and consumption over the ex-colonial, developing lands. In the planet at large, a wealthy "North" in stark contrast to a still-unmodernized "South" transferred Disraeli's dictum to the world community.

The cities once again took the brunt of this contrast—in the pockets of poverty in slums and ghettos in developed lands, in the shantytowns and *favellas,* the *calampas,* and *bidonvilles* which began to surround developing cities with the squalid camps and shelters of millions of migrants from rural misery.

Even where the boom brought rapidly rising and widely diffused prosperity, the pressures increased. A new surge of suburban growth followed the increasing ownership of automobiles. City after city exploded into the surrounding countryside, extending with them the concrete expanses of our vast highways and modern conurbations. Metropolitan areas which had reached a million inhabitants before the First World War went on to grow into conurbations of two and six and seven million.

Moreover, the sheer scale of effluents to be disposed of and materials to be junked—1900 pounds per person per year for instance in the United States—coupled with the wide geographical spread of urban and suburban settlements meant a sudden, vast increase in the strain on air and rivers which, by and large, continued to be used as "free goods" still able, without cost, to dispose of the unlimited wastes of the millions upon millions of new high-consuming citizens. Nor has this burden simply proved to be an intensification of earlier pollution. It has also changed in character. The very freedom of recently invented technologies to devise, manipulate, and create new substances in the wake of chemical discovery has added a still-unmeasured element of risk to the effluents of massively expanding industrial systems.

But the really new risk, foreseen in the past by only a few of the world's economists, is that such a pressure of rising demand may begin to put intolerable strains on what had appeared to be the planet's limitless resources. This risk had been masked in the thrusting nineteenth century by the opening up of all the planet's temperate lands to European settlement and later by the extraordinary productivity of new forms of energy and chemical transformations. Nor did anyone before 1945 fully foresee the runaway growth in world population.

But once the risks of future shortage begin to appear, the insufficiency of market mechanisms to deal fully with them becomes equally obvious. The market has only one answer to scarcity—to put up the price. In the longer run, this may encourage the invention of alternative technologies which economize materials and energy. But in the short run price rises are usually unavoidable and this fact leads not simply to economic but to political problems. If costs and prices rise, something has to give, either private standards or public spending—or the planet's integrity. For contemporary political systems this dilemma has an extra edge of difficulty in that public spending tends to include a large element of the expenditures which are least useful to consumers and most inflationary in their effect—the cost of armaments. And this is simply one more reminder that, decisive as the market has been as a stimulus to the development of our disturbed, fragmented, and powerful modern order, a far more potent agent of change has been Bacon's other idol, "the idol of the tribe," the emergent nation-state.

Nations and Markets

This third element in the powerful trinity of forces—science, the market, the nation—has its roots in concepts of political sovereignty which in one form or another are as old as the hunting clans. The important point is that the cohesive modern nation-state has developed the authority, the organization, the will, and the energy to do three critical things.

First of all, it created an internal market, wide yet coherent enough to launch the Industrial Revolution. A nation the size of Britain or France, for example, proved to be a very promising area for commercial organization. It was large enough to create a "critical mass" of demand for rising production but also small enough to be effectively administered and governed. In large, rambling territories like Central Europe with its ill-defined Holy Roman Empire, surcharges and duties would be charged at every toll bridge and city gate. In China, too, distances and local obstacles were too great for the quick emergence of a truly national market. It is significant that of all the nations of Asia, it was in Japan, an ancient island sovereignty most resembling Britain, that nationalism and industry began, in the nineteenth century, to go hand in hand.

Next, the nation states of Western Europe created, without directly intending it, a world-wide market. Europe's maritime powers—Portugal, Holland, Britain, France—first sought trade with Asia and then tried to keep all rivals out precisely because Europe was so short of the consumer goods such as silks and spices men most valued in the past. In the eighteenth and nineteenth centuries, Britain, outdistancing other trading nations, created the first genuinely world-wide system of exchange, exporting African slaves to the New World to produce sugar and later cotton, exchanging the sugar in Europe for iron ore and timber and nautical stores, using the ships to trade for cloth and spice in Asia and paying the Asians with bullion taken from Africa (as the name of the British gold coin *guinea* implies).

This vast entrepôt trade grew more profitable when the collapse of internal authority in India established Britain's colonial control (exercised significantly through a semipublic trading corporation, the East India Company). This takeover permitted the British to phase out Indian hand-loom textiles and substitute Lancashire cloth first at home and then for export—the first modern case of successful import substitution and one which may well have given Britain's textile industry the elbowroom needed to invent full-scale factory production.

This imperial expansion and world-wide commercial predominance was not in any way part of a preconceived plan. It was the very quintessence of the unintended order. But once launched, the type of commercial relations established by Britain and copied by its industrializing

Atlantic neighbors proved remarkably persistent. Basically it involved the exchange of capital and manufactures from developed lands for raw materials produced in developing and often colonial countries. Much the same pattern persists today in the economic relations between developed and developing nations.

The third impact on man's environment flowing from the expansion of national power brought science and state together in the pursuit of war. From the sixteenth to the nineteenth century, the oceans of the world were full of brawling, struggling European rivals, chasing and fighting each other for the control of goods, monopolies, and trading posts. One consequence of this maritime rivalry was to stimulate research into metals light enough to produce cannons which would not sink the ships which had to carry them. Some of the most skillful technical innovators of the eighteenth century owed their head start to naval and military operations. In Britain, for instance, Henry Cort began life as a naval agent and went on to revolutionize iron production. Henry Maudsley, happily remembered for the invention of the beer handle, began his career in machine tools in Woolwich Arsenal.

But it was not only colonial rivalries that powered the furnaces of war. Once Germany, Russia, and Japan belatedly joined the ranks of self-consciously nationalist states and sought to assert their sovereignty and to expand their interests in a world largely controlled by the earlier established nations, nationalist rivalries exploded into a series of wars which in the twentieth century twice engulfed the planet and still leave behind such powerful legacies of fear and distrust that they inspire a general spending on armaments on the order of $200 billion a year.

If the modest wars of the seventeenth and eighteenth centuries sparked a genuine acceleration in technological skills, it can be imagined what the twentieth century's global wars and global arms have brought about in the way of thrusting, single-minded technical discovery and applied research. It was the nineteenth century demand for accurate small arms that encouraged the industrial evolution of interchangeable parts and the high-speed lathes, new drilling equipment, hardened metals, and new alloys needed to make them. All these inventions have proved critical in the development of mass-produced modern "consumer durables," above all the motorcar.

The First World War gave agriculture the tractor via the tank and put a booster of incalculable force under the automobile and the airplane.

The Second World War speeded up by decades the evolution of all kinds of electronic equipment and ushered in the final Promethean theft—the discovery that the atom's nucleus could be split and its earth-destroying power used for human purposes. And all this war-inspired stimulus, all this investment, all this input of trained minds and infinitely elaborate equipment into weapons has served the most negative, the most divisive, the most wasteful set of purposes that mankind can set itself—and at an incalculable price in welfare and resources.

In short, mankind has still found no organized system for reconciling the driving demands and ambitions of national statehood with the wider unities of a shared planet. With all the growth, all the enrichment, all the apparent market success of the fifties and sixties, men are left in deep unease about the current condition of their planet.

Just as the grouping together of factories in urban concentrations during the last century vastly increased the pressures of filth and contamination, in our modern mass economy the sheer concentration of consumers is beginning to show a comparable effect. Drainage systems, waste disposals, urban structures—none was designed for such a flood of disposable artifacts bought, enjoyed, and thrown out by such a flood of people.

Above all, the distribution of prosperity is dangerously skewed. Within each affluent economy, minorities who are handicapped by ethnic prejudice or age or sickness tend to be left behind to observe vicariously on television how the luckier three-quarters live. And in planetary society as a whole, it is three-quarters who live badly and, as their numbers rise, face bleak prospects of living better.

To restore balance and hope, to moderate the despairs and pressures, to achieve common policies for a viable political order are thus the preconditions of any decent human environment on planet Earth.

If all man can offer to the decades ahead is the same combination of scientific drive, economic cupidity, and national arrogance, then we cannot rate very highly the chances of reaching the year 2000 with our planet still functioning safely and our humanity securely preserved.

Equally, however, a careful recognition of where our failures lie is the first step toward the wisdom and restraint our overwhelming power requires of us. In every case, the needed steps take us away from division, from single-shot interventions, separatist tendencies, and driving ambitions and greeds. We have to grasp and foster more fully the truly integrative aspects of science. We have to revise our economic manage-ment—of incomes, of environments, of cities. We have to place what is valuable in nationalism within the framework of a political world order that is morally and socially responsible as well as physically one. Our errors point to our cures and on the basis of man's survival up to this point, it is not wholly irrational to believe that he can learn from his mistakes.

Part Two: The Unities of Science

3 ENERGY AND MATTER

WE MUST BEGIN OUR SEARCH FOR countervailing insights in the first place in a wider philosophy of environmental reality. There is a profound paradox in the fact that four centuries of intense scientific work, focused on the dissection of the seamless web of existence and resulting in ever more precise but highly specialized knowledge, has led to a new and unexpected vision of the total unity, continuity, and interdependence of the entire cosmos. By looking beyond the bewildering confusion of superficial events and plunging into the depths and details of measured structure and energy, scientists have created a unitary picture of the physical and living worlds far more radical than any prescientific thinker could ever have conceived. To see scientific method as primarily a dissecting, dividing force is therefore to miss the astonishing breakthroughs to unitary knowledge that have occurred in the twentieth century. At their center lies man's increasing grasp on the central cosmic fact of nuclear energy.

From independent efforts of research in a number of countries there has emerged in the last century a new picture of the nature of the atom. A young Danish scientist, Niels Bohr, developed the idea that it might resemble a tiny solar system with the nucleus as the sun at its center. Around this core negatively charged particles or electrons, weighing no more than 1/2000 of the whole atom, whizz in fixed orbits, like planets round the sun, their negative charges holding them away both from the central positive nucleus and, in atoms with numerous electrons, from each other as well.

This view of atomic structure set the stage for a remarkable reinterpretation of Mendeleev's table of elements. The order of these elements, which had been first established by empirical estimates based upon comparisons of their relative atomic weight, came to be seen as an orderly progression based on the electrical charge of the central nucleus and the number of electrons orbiting around it. As the charge and the number of electrons increases, so does the weight. So also does the complexity of the orbits. Electrons cannot move at random round the nucleus. The orbits are fixed. But they can, as it were, jump up and down stairs within the atom. Energy is released when they fall from an outer to an inner orbit. When a suitable charge of energy is introduced into the system from outside, they can jump up an orbit and even right out of the atom.

This picture in fact gives too undynamic a picture of the atom's behavior. One can compare the electrons' orbits to a spinning cocoon of movement round the central nucleus, each electron making over a hundred million billion circuits a second. The apparent stability of the structure springs from the sheer speed, number, and everywhere-at-onceness of the circuits.

At the heart of the atom sits the formidable mystery of power, the nucleus. It is unbelievably small. If an atom were expanded to the size of a house, the nucleus, to scale, would still be no bigger than a pinhead. Yet the nucleus, with its strong positive charge, accounts for nearly 100 per cent of the atom's weight. Each element has precisely equal numbers of protons and electrons. Hydrogen, with one proton in the nucleus and one electron orbiting around it, is the lightest. Uranium, with ninety-two protons and ninety-two electrons, is the heaviest element not made by man. But some elements with precisely the same basic balance of protons in the nucleus and electrons outside it come, surprisingly, in different weights. It was only in the last half-century that scientists were able first to postulate and then discover another particle within the nucleus—the neutron. It has no electrical charge at all but its presence and numbers explain why elements come in different weights. Their varieties have come to be called isotopes. Uranium 238, for instance, has 92 protons, 92 electrons, but 146 neutrons. Uranium 235, the fuel of the first atomic bomb, has 143 neutrons.

But the neutron is not the only particle the researchers discovered inside the nucleus. As the atom smashers became more and more powerful and atoms were dashed with increasing violence against their targets, out of the nucleus poured not only neutrons, but showers of even tinier particles. Some of them, unimaginably minute as they are, can apparently combine with one another for as little as a hundred-thousandth of a billionth of a second. Known as resonances, they are the furthest reach in definable human experience of the indefinitely and infinitely small. But this is not the end. The search for the ultimate components of matter continues.

This heroic voyage of atomic discovery revealed that the behavior of the minute building blocks of matter underlies all the various kinds of energy. Thermal energy makes the electrons in various elements oscillate more and more violently until they burst out of their shells, ready to enter other combinations. Chemical energy is released when electrons from the shell of one element jump into the shell of another element to make a more stable compound. Electrical current is created when electrons, pulled out of their shells by electromagnetic attraction, begin bouncing from atom to atom, along a good conductor like a copper wire.

All energies have an element of "flow" in them. Electrons move from hot to cold, from more unstable to less unstable elements, from higher to lower levels in response to electromagnetic forces. The source of virtually all the energy on our own planet, in the whole solar system, and probably in the whole cosmos is explicable in terms of atomic action similar to the process at work at the core of the sun. And it is here that man has learned what is for him the ultimate formula of physical reality.

Not only are all forms of energy interchangeable. In the last analysis, energy and matter are interchangeable, too. The material universe of which we ourselves, our peoples, and our planets are phantasmal parts is made up of a single sweeping, pulsing force of energy, existing for eons in the spinning galaxies of the firmament, for minute flashes of time in our physical bodies and questing brains—but all parts of a common throbbing impulse of unimaginable force. The scientist's vision of orderly law, the philosopher's dream of overarching unity, the sage's sense of a single cooperative order are not baseless imaginings. They are facts, not dreams. The primal alphabets of being all show forth a single unified system of interrelated energy and law.

To understand how in our own day matter has been turned into energy, we have to go back to the balance between positively charged protons inside the nucleus and negative electrons orbiting outside it. This balance explains the relative stability of the atom. But how on earth does the nucleus itself hold together with only positive protons and neutral neutrons? The protons, with no opposing force, ought to be pushing violently apart. But they do not. Physicists have therefore concluded that a super-strong bond or nuclear force keeps the nucleus together. When certain nuclei are split, the energy given off represents, in fact, one of nature's mightiest releases of energy.

One possibly helpful analogy is to compare a nucleus with a screen door with a tremendously powerful spring. You have to pull and pull with great effort to get the door open. Then it swings to with a mighty slam. This super-slam is the energy releasd when the protons and neutrons that have been pulled apart come together in another nucleus. This energy is the basis of nuclear energy produced either by fission or by fusion. It is also the energy which under the pressure of war, fueled the first atomic bomb.

This first bomb was based on fission. Enough neutrons bombarded plutonium, a man-made element derived from uranium, to set in motion an explosive chain reaction, as released particles bombarded further atoms, which in turn bombarded still others. This was the beginning. Nearly ten years after the first fission bomb, the energy derived from fission was used as a trigger for another even more violent bomb based upon a reaction of a kind which occurs in nature, in other words, the thermonuclear process, the fusion of hydrogen nuclei which takes place at the heart of our sun and in all the stars of all the galaxies.

We are accustomed in our earthly environment to think of physical reality in three states—liquid, solid, or gaseous. But there is a fourth state to which physicists have given the name "plasma" (with a meaning entirely different from that of the word in biology). This designates the state of matter when it is subjected to temperatures above and beyond 5000°F and it may be that hydrogen plasma is the original stuff of the whole cosmos. We know from modern astronomy's 200-inch lenses and radio telescopes that outside our own galaxy, the Milky Way, there are at the present limit of our telescopes 100,000 million more such galaxies, each with 100,000 million other "suns." Inside the whirling galaxies, centers of high density begin to form and to contract round a gravitational center which is the birthplace of a star. As the atoms fall inward and collide with each other under the force of gravity, heat begins to build up and within the seething plasma—whose temperatures are rising to equal those of the sun, a fantastic 25 million degrees Fahrenheit— hydrogen ions (or nuclei) fuse with others to form the element helium. This fusion releases an enormous flood of energy, which further stokes the nuclear fires and sets going a chain reaction at the core of the star.

Our star, the familiar sun, transforms by this process 657 million tons of hydrogen into 653 million tons of helium every second. The lost four million tons of mass are flung off as energy which radiates through the whole solar system. Thus is Einstein's famous equation $E = mc^2$—the energy of an object equals its mass multiplied by the speed of light squared—worked out for us daily in the powering of our solar system. Of this energy, planet Earth receives only one two billionth part, of which we use less than 1 per cent. We can, it is clear, stand only so much of the sun's vast radiance.

Yet we have already repeated on earth the process which flings it forth. The fusion thermonuclear bomb—the so-called hydrogen bomb—was first exploded in 1952 by using a fission trigger of uranium 235 to fire a flash of heat so intense that it could fuse hydrogen into helium, thereby releasing the kind of energy which is generated by the sun. All other environmental risks, it is clear, fade into insignificance compared with the possibility that this terrible weapon might ever come to be used in all-out war.

Since that time, the "milder" process of nuclear fission has, it is true, been harnessed for peaceful use. Given the interchangeability of all forms of energy, atomic power can be used to generate electricity when neutrons bombarding the nuclear fuel—U235 or plutonium—build up chain reactions which heat water to the temperature of steam, which then undertakes the familiar tasks it has performed for decades in coal- or oil-fired generating stations—turning the generators that move the magnetic fields that send the electrons scurrying along the power lines. The difference lies in the nature of the fuel. Nuclear energy in any mishap is an unseen killer. Moreover, uranium debris can last for millennia, still giving off dangerous radioactivity. None of man's inventions of energy have been foolproof. The earliest steam engines blew up in men's faces. One of the first steam engines ran over a cabinet minister. But we have crossed a new frontier of risk when using radioactive materials which can blast genes and linger on for centuries and we have done so just as we have acquired a new vision of the fragility of living things and of the very special conditions which had to prevail in our planetary environment before life could first come into being and then persist and evolve.

It is a strange coincidence. Man has lived for perhaps a hundred millennia on planet Earth with no knowledge of how such a place, such a system, such an environment came to be formed in the first place. It is only in the last century that anything resembling a picture of his planet's birth and growth has been put together for him. It is still being rethought and revised. And just as he begins at last to understand its chanciness, the hazards it has survived and the hair-raising accidents it has managed to avoid, he comes to conquer for himself the blinding, death-dealing energies of the sun which, for millennia, represented the greatest threat to the emergence of life. Creator, destroyer, source of all energy, potential source of annihilation—it can operate without danger on planet Earth only through a whole series of delicate and complex mechanisms and defenses which have taken billions of years to build up. We cannot understand the full meaning of our planetary condition, of the environmental necessities required for human survival, without some knowledge of both the creative and destructive relationship between the nuclear fusion at the sun's core and the emergence and preservation of life on earth.

4 THE ALPHABET OF TIME

THE FUNDAMENTAL POINT in the relationship is, quite simply, the ability of the planet to develop in slow stages mechanisms which protect it from the destructiveness of solar radiation, yet enable it to use its life-giving energy. Billennia after billennia, there poured down on the earth through unimpeded space a whole spectrum of solar radiations up to frequencies of 10^{22} cycles per second capable of destroying any form of life. But progressively defenses and mediations arose to permit the emergence on a lifeless planet of the covering of living things for which the Soviet physicist Vladimir Ivanovitch Vernadsky invented the word *biosphere*.

The first protection came from water. In the searing heat of the still-molten globe, water, which turns to vapor at $212°F$, hung round the earth in layer on layer of impenetrable cloud. Underneath, as the earth's heat receded at uneven speeds, its core remained molten while exterior crusts crumpled and folded and tore apart in tremendous gulfs and upthrusts of rock. As the process of cooling went on, the cloud above turned from vapor to water. The rains began to fall. They fell for years, for centuries, for millennia in a continuous, global downpour, filling the crevices and the gulfs. They covered the lower lands. They climbed up the mountains. They all but filled the Southern Hemisphere. The oceans were born and became the cradle of life.

The next stage of development follows the ending of the rains, some three billion years ago. As the water sloshed about on the earth's unstable and volcanic surface, the crumbling and eroding rock brought down a "soup" of chemicals into the oceans. When played on by electrical discharges and by the sun's relentless radiation, these chemicals began to form complex molecules. Carbon, with its four bonding points, was particularly receptive and today there are over two thousand organic compounds containing carbon. There is no life without it.

In recent decades, scientists have bombarded chemical mixtures having some similarities with the assumed composition of the oceans' primitive "soup" and have thus produced a few of the organic molecules found in the building blocks of life. But, just as the origins of life remain mysterious, so does its foward movement. What is certain is that life might have remained at a very primitive stage if a new kind of shield had not begun to build up. The starting point was the release from the planet of a protective atmosphere, containing oxygen and ozone, which intervened between the waters and the sun's lethal radiations. Below this shield, a new life-expanding process, photosynthesis, began to enable living things—bacteria, algae—to use the sun's radiance for the creation of organic matter and for the release of more oxygen.

Photosynthesis is, at bottom, an incredibly elaborate transformation of light energy into carbohydrates (or sugar), which is the food of all living things. In this process within the bacteria, chlorophyll—a molecule made of carbon, hydrogen, magnesium, and nitrogen—releases energy when it is struck by the sun's rays. With this energy, it absorbs and breaks up water molecules, combines their hydrogen with carbon and other chemicals to produce sugar and lets the oxygen off into the atmosphere. The process is much more elaborate than the workings of a modern petrochemical complex. It takes place in a group of cells not a billionth of an inch across.

At the same time, in a balancing process of respiration, the tiny sea plants began to pull in oxygen, push out carbon dioxide, and produce, as an end product, water and usable energy. Thus the substances which were drawn in at the beginning of photosynthesis, water and carbon, are released at the end of it, and the oxygen which had been released is reabsorbed. On these two great cycles—the carbon cycle and the oxygen cycle, together with the lesser inputs of nitrogen, sulphur, and phosphorus—depends the whole breathing life of Earth's plants and animals.

The living cells capable of such prodigies of chemical transformation had obviously evolved far beyond the simple fission practiced by the type of primitive bacteria fossilized in the so-called Figtree formation in South Africa some three billion years ago. A billion years later, fossils representing their successors turn up in eastern California. Among them were algaelike organisms now giving evidence of a nucleus in the core of the cell. For the next billion years, in warm, sheltered coastal waters and estuaries they increased, through photosynthesis, the release of oxygen —today, a full quarter of all the oxygen we breathe is produced by the infinitesimal phytoplankton lying on the seas precisely at the point where air and water meet—and where, all too often, mammoth tankers empty their oily bilge.

Living cells, meanwhile, were evolving by responding to the new opportunities. The finding of a new fossil, the so-called spriggina, in the Ediacara Hills in Australia, indicates that some 700 million years ago there were living things with more than one cell—the first evidence of the metazoa, a class to which all elaborate organisms belong. Again we do not know how the greater complexities of organization began and evolved. Indeed, only in the last twenty years have we learned how the genes, through the double helix of their DNA (deoxyribonucleic acid), transfer to cells precise instructions for their organic life. This is just as true for man as for the simplest bacteria. The human body consists of some 600,000 billion cells which reproduce themselves, interact, fend off insults, and respond to the opportunities of their daily environment under the precise instructions of their own DNA.

The first primitive cells probably evolved under the protection of water, at a time when volcanoes and earthquakes were still shaking the earth, still sloshing the oceans over the land and back into the abyss. But this random process is precisely what seems to have prepared the next great surge of life. Eventually some sea plants contrived to take to the rock. The most primitive land plant, the Cooksonia, is found in a fossil dating from about 450 million years. More and more plants covered the rocky surfaces and animals probably followed. Fish became amphibians with leglike fins to help them stump across the sea-washed mud of lakes and estuaries. Gills turned to lungs as they learned to breathe oxygen. About 350 million years ago the landward movement of living things becomes a flood. Plants begin to cover the rocky surfaces of the earth with the new green of breathing leaves and to set in motion the planet-wide processes of photosynthesis and transpiration. Some three-quarters of the atmosphere's essential oxygen came to be recycled through the plants, providing breathable air for all the planet's creatures.

As vegetation spread to hot equatorial regions or to the more temperate zones of the north and the south, trees and plants adapted themselves to the changing climates and began to form what are now called biomes

—such typical patterns of vegetation as the evergreens of the north, the eucalyptus of Australia, the palm trees of the tropics. The roots of plants, progressively disintegrating the rock, added to the erosion of the ages and helped to build up the thin and precious envelope of soil which sustains all plant growth and hence all forms of life. Some types of soil bacteria fixed the air's nitrogen, which provided nutrients for the plants and contributed to the rungs of the DNA ladder in all living things. The circulation of nitrogen through air, soil, and living things is but one example of the many natural cycles of elements which are essential to the economy of the biosphere. In fact, all nutrients continuously recirculate through the planet's natural divisions—atmosphere or air, the hydrosphere or the waters, the lithosphere, which is the rock.

Wherever air and water and rock meet, living things establish their home. The biosphere goes not much higher than 17,000 feet into the air. It rarely reaches much further than 10,000 feet in the ocean. Apart from deep drilled mines and wells, the usable part of the lithosphere is represented by a few feet of soil. But this relatively tiny fraction of the planet can support life only by engaging in continuous interchanges between its three vaster partners.

All living things have to be adapted to their surroundings in order to survive and reproduce their kind. Natural selection is the fundamental mechanism of this adaptation. But this phrase gives little idea of the infinite variety of stratagems by which living things come to occupy different niches and to produce the incredible variety of shapes, colors, movements, patterns of courtship, of escape, and challenge which make up the richness of the biosphere.

Natural selection involves, of course, conflicts for limited amounts of food and space. Competition in this sense exists all over nature. But it is not so sheerly ferocious as some nineteenth-century thinkers made it out to be after the publication of Charles Darwin's *On the Origin of Species*. Prudence, cooperation, indifference, parasitism all play a part. Groups of animals may act together to protect each other. One thinks of the little circle of quails sitting tail to tail at night ready to fly off together in a bomblike explosion of squawks and feathers at the least sign of danger. In the herd or pack, animals rarely attack each other under natural conditions and often evolve common techniques for securing food or protecting each other.

Stable environmental relationships imply an interconnected variety of food chains and food webs, which contribute, as it were, the grids of energy upon which survival depends. A typical food chain is a sort of pyramid. At the bottom are the plants, which use minerals from the soil and energy from sunlight to produce their own tissues. Then come the herbivorous animals, which feed on the plants. Then the flesh-eating animals or carnivores, which are fewer in number than the herbivores. Finally there is man, the most successful hunter (or predator) of them all.

From the original emergence of plant and animal down to the present day, this energy chain of food-giving and food-getting has not changed in its essentials. A typical chain in a forest comprises a multitude of nuts falling from the trees and feeding a smaller multitude of squirrels; they in turn are eaten by a much smaller number of foxes; at the head of the chain the human hunter now shoots the fox, which he once ate. The excreta from the animals fall to the forest floor, feed microbes which create the humus from which the trees grow which produce the nuts. Forests have a great variety of such chains, and, in a mature forest—a so-called climax ecosystem—all the various food chains are self-sustaining. In theory, the climax ecosystem can thus live its life of gently rapacious vitality for millennia.

The forest is only an illustration of one of nature's infinitely various food chains, each with its own composition and complexity. Chains interconnect with each other to form webs which include the widest variety of plants and animals. Some chains can extend across continents, for example through the birds which are part of them. There is still some uncertainty as to the mode of transport of DDT, but it is a striking manifestation of the interrelatedness of things on a global scale that this insecticide, used in temperate and tropical countries, has turned up in the fatty tissue of penguins.

Although ecosystems can be extremely stable, many are vulnerable. Ecological balance does exist in a climax area, but one violent windstorm or volcanic eruption can destroy in a matter of minutes the equilibrium of centuries. Moreover, the disturbance need not be on this shattering scale. The sudden removal of one small component in a food chain may cause others to starve. The ecology of a freshwater trout stream running clear through woods and meadows can be disturbed by an insecticide that kills the fishes' standard diet, or an effluent that stimulates the algal growth. As a result, the rainbow trout vanish, the mudfish remain.

Delicate flowers bloom no longer, only reeds and sedge.

An extreme form of ecological shock occurs when there is a breakdown of the natural mechanisms built into nature for self-protection and self-renewal. Then the lemmings surge to uncontrollable numbers, water hyacinths invade and choke the reservoirs and waterways, rabbits nibble the Australian sheep out of grass, coypus break loose and burrow through dams and drainage ditches. Again and again, this type of upheaval follows the sudden introduction of some new factor—biological species or chemical—which has no established links with existing patterns.

Long before man's hominid predecessor brought to bear on the earth skilled hands, a versatile body, and an inquiring mind, the natural world was already incredibly complex and rich in animal and plant species with their songs, colors, scents—but also with their pitfalls and challenges. This is the complex kind of world in which there occurred, approximately 100,000 years ago, the unexplained and unparalleled enlargement of the human brain, which resulted in *Homo sapiens,* thus bringing into play on earth a type of force different in kind from other natural forces, a creature within the natural system but capable of seeing his place within it and even entertaining the illusion that he could manipulate, command, and conquer it wholly for his own designs.

5 A DELICATE BALANCE

AT THIS POINT let us look back again over the vast panorama of space
and time that has unfolded for us as we have attempted to recognize the
planet's underlying systems. Its foundation is a most majestic unity.
Matter and energy are simply different aspects of the same fundamental
reality and in all their manifestations obey ineluctable cosmic laws.
Furthermore the operation of these laws through all the infinite varieties
of material things and energies generates another kind of unity—the
dynamic equilibrium of biological forces held in position by checks and
balances of a most delicate sort.

There exists a single unified system from one end of the cosmos to
the other; in the last analysis, everything is energy. Its larger spirals are
the galaxies, its smaller eddies suns and planets, its softest movement the
atom and the gene. Under all forms of matter and manifestations of life
there beats the unity of energy operating according to Einstein's law. Yet
this unified stuff of existence not only twists itself into the incredible
variety of material things; it can also produce living patterns of ever-
greater complexity—from the gas bubble in the original plasma to the
single cell, from the single cell to the crowning complexity of the human
brain, in which billions of neutrons, occupying only one-twelfth of the
cortex or forebrain, may each have as many as 270,000 means of direct
communication with its neighbors.

Within our own planet the interplay between vast cosmic unities and
the minute instruments of equilibrium is the very stuff of existence. We
know that the energy of the sun is poured out in almost limitless bounty.
But we know, too, that the intermediaries and products of all this bounty
—the leaves, the bacteria—are far from limitless. Remove the green
cover from the soil of Central Africa and it becomes a brick-hard,

everlasting laterite. Cut down the forests, overgraze the grass and productive land turns to desert. Overload the waters with sewage or nutrients and algae consume its oxygen, fish die and produce stinking gas as they decompose.

It is because there are so many potential paths toward points of irreversible no return that the self-repairing cycles underlying all living systems—the unities of dynamic balance we call ecosystems—cannot survive indefinite overloading or mistreatment. Admittedly, the regenerative powers of life are astonishing. Living things have survived the glaciations, the volcanic convulsions, the earthquakes, typhoons, and tidal waves that have torn through our unstable planet over the billennia. But the warning is there. Like the giant reptiles of the Jurassic age, some species have gone the "way to dusty death."

And this brings us to a final balance. The intimate, inescapable interdependence of living things implies a certain stability, a certain dynamic reciprocity. Its weakening or destruction unleashes the capacity of creatures to destroy each other and themselves as well. There exist in nature many different patterns for ensuring to a species its food and young. Most of these patterns entail the eating of one species by another, all the way up the food chains of nature. Some animals demonstrate high though specific and limited degrees of cooperation—within the nest, the lair, the pack, the herd. Some exhibit a host-parasite relationship which can express a subtle equilibrium even though it appears unattractive to human eyes. But behind the interrelationships lies the risk of unpredictable and sometimes destructive consequences if the delicate equilibrium is overturned. A new species introduced, a chemical balance upset, an island erupting into the sky, the slow onset of the ice—all such disturbances can elicit so violent a response that the system may not be capable of returning, by itself, to a desirable and stable system.

If these are indeed the lessons learned in piecing together the infinite history of our universe and of planet Earth, they teach surely one thing above all—a need for extreme caution, a sense of the appalling vastness and complexity of the forces that can be unleashed, and of the egg-shell delicacy of the arrangements that can be upset. And this clearly has significance for more than strictly scientific knowledge. It is not only the new science and the technologies derived from it that dominate the life of modern man. His society is also driven forward by an unparalleled intensity of material wants and by still-unconquered drives of national separatism and power. The question has therefore to be asked how, in an age dominated as never before by separate nationalist aspirations and pretensions and by the promise of indefinitely rising material standards, is the new moderation to be sought? How can cogent arguments be marshaled for balance, for cooperation, for that awareness of reality which all the great sages of mankind, without exception, have held to be the root of human wisdom and hence of human survival? If their witness has been so very largely in vain, how can we hope now for better insights and better will?

There is, however, something clarifying and irresistible in plain scientific fact. The astonishing thing about our deepened understanding of reality over the last four or five decades is the degree to which it confirms and reinforces so many of the older moral insights of man. The philosophers told us we were one, part of a greater unity which transcends our local drives and needs. They told us that all living things are held together in a most intricate web of interdependence. They told us that aggression and violence, blindly breaking down the delicate relationships of existence, could lead to destruction and death. These were, if you like, intuitions drawn in the main from the study of human societies and behavior. What we now learn is that they are factual descriptions of the way in which our universe actually works.

Both the development of atomic science and the piecing together of the planet's and of man's evolution—master intellectual achievements of modern times—have provided a solid basis for a completely new appreciation of the unity, interdependence, and precariousness of the human condition. And since this reality comes to us with all the weight of scientific proof and cogency, we can hope that it will be more convincing than was the earlier, less scientifically substantiated knowledge. The unraveling of atomic structure and the unfolding of our biological history thus offer a remarkable paradox. On the one hand, nuclear power gives man the means of self-annihilation. On the other, the delicacy and balance of the evolutionary process offers him the perspective he needs to avoid planetary suicide.

This warning and this hope echo in the ancient wisdom of all the earth's great cultures. But perhaps because Western man has been largely responsible for opening up the furnaces of nuclear power and for penetrating to the most intimate mechanism of life, it is in Western tradition that we find the most urgent warnings against arrogant and unheeding power. For the Greeks it is Prometheus, stealer of fire, who is chained to the rock. Nemesis in the shape of shrieking, destroying harpies follows the footsteps of the overmighty. In the Bible, it is the proud who are put down from their seats; the exalted are those of humble spirit. At the very beginnings of the scientific age, in the Faustian legend, it is the man of science who sells his soul to secure all knowledge and all power.

All this does not imply a retreat from the fantastic achievements of science or skepticism concerning its immense possibilities for future use. The scientific method employed now to decipher not the separation but the interconnections of material things can provide men with better, more reliable, and wiser means of working with his environment. Nature has so many unstable, unpredictable, and violent facets that man needs all his probing intelligence and enormous potentiality for understanding to enhance and stabilize its capricious bounties. In fact, in Goethe's retelling of the Faustian legend, the end is not torment but redemption —when Faust at last uses his powers not for aggrandizement but to drain a marsh and feed the people.

But the warning remains. Powers on such a scale require the furthest reach of wisdom, detachment, and human respect in their exercise. If man continues to let his behavior be dominated by separation, antagonism, and greed, he will destroy the delicate balances of his planetary environment. And if they were once destroyed, there would be no more life for him.

Part Three: The Problems of High Technology

6 THE DISCONTINUITIES OF DEVELOPMENT

WE TURN NOW TO the most urgent economic problems raised by the environmental issue, by the effect of high demand, consumer pressure, resource use, and the particular problems of dealing with the increasing concentration of people living in heavily built-up areas. Policies which, as in the past, tend to leave to haphazard decision such questions as external diseconomies or rising shortages or the development of urban regions are less likely than ever to produce satisfactory solutions now that numbers, urbanization, and the impact of technology are all increasing and interacting at unprecedented speeds. Just as scientific thought has discovered the need to underline its preoccupation with the integrative aspects of reality, so, too, in economic life the interdependence and complex interactions of man's industrial and urban life have now to receive a larger emphasis.

But before we look at the first issue—that of the often ununderstood and unpaid-for diseconomies imposed by the modern technological system—we have to determine our general approach to the new planetary economy. In a sense, it is a continuum. No nation in the world is outside the network of trade and investment which has been spun round the globe in the last three centuries. None is beyond the reach of modern instantaneous communication. Fashions, diets, schooling are all influenced by world-wide trends. All nations are in varying degrees involved in the new technological order.

Nor do problems and difficulties vary completely from region to region. Air pollution can be as great a problem in Seoul as in Chicago. Unmanageable wastes can pile up in Bangkok as well as in Manchester. Moreover, existing areas of very high industrial concentration, pollution, and difficulty may simply be demonstrating in chilling fashion the ultimate fate of all the world's peoples as they enter more fully into the industrial order. There would seem therefore to be a strong argument for treating the planetary economy as a continuum and studying the impact of pollution, of urbanization, of resource uses and shortages on *all* economies, irrespective of their condition—pretechnological, newly industrial, or moving on to degrees of urban and industrial concentration which presage wholly new complexities of pessure and congestion.

Yet there are a number of cogent reasons for looking at the problems of relatively high-income and low-income countries separately. The separation has nothing to do with judgments about what is valuable or creative in national life. Low-income countries include some of the greatest and most ancient of the world's cultures, and some of its oldest and most continuous political states. The need for separation stems from certain sharp contrasts between conditions and opportunities in what, for want of a better definition, have come to be called developed and developing states.

Taking a fairly arbitrary level of national income—say, $500 per capita—as a very rough dividing line, we can say that the countries below this level tend to share certain acute problems which directly affect their environmental outlook and which are not experienced on the same scale —or at all—in developed lands.

Their populations are growing almost twice as fast as those of wealthier countries and twice as fast as did industrializing states in the nineteenth century.

Their soils tend to be more fragile and their climates less reliable and moderate than in temperate zones, where the majority of high-income states are to be found. Moreover, much less is known in scientific terms about specific natural conditions and needs in developing lands.

With a few exceptions, urbanization has come ahead of industrialization and while this fact exposes a country to some of the traditional economic diseconomies of overloaded air and water, it also creates the risk of far greater *social* diseconomies than any known during the process of nineteenth-century industrialization. With a work force growing at twice the traditional speed and a rising tendency, in farms and factories, to invest in capital-intensive equipment, endemic worklessness is added to the more traditional evils of poor housing and sanitation and lack of proper training and skills.

All these pressures create environmental problems, both economic and social, which are sufficiently different to merit consideration apart from the issues which are the most urgent concern of high-income lands. However, the distinction remains arbitrary. Rich and poor, developed and developing, industrialized and pretechnological—all are enmeshed in myriad webs of trade, communication, and influence, all are struggling to adapt the technological order to truly human ends, all are involved in the welfare and survival of their fellow communities, all must inescapably share a single, vulnerable biosphere. Whatever their immediate differences, the environmental issue confronts them all with an ultimate challenge—the survival and good estate of their planetary home.

7 THE PRICE OF POLLUTION

Market Costs and Social Costs

IN DEVELOPED ECONOMIES TODAY there are, above all, three areas of production and consumption, in which satisfying the myriad unpredictable and fluctuating desires of modern man can come into conflict with basic social and environmental needs, where market signals, designed to give some manageability to all the diversity of demand, provide less than satisfactory answers to society's environmental problems and where, as a result, new policies, stressing the wider contexts of order and amenity, have to be introduced.

These areas are, first, the problem of production costs which disregard external diseconomies, second, the pressure of modern urbanization, third, the growing risk of scarcities—in materials, in energy—as a result of continued economic growth.

Our understanding of external diseconomies is inherited from an earlier industrial tradition and, today, in virtually all developed economies, whether centrally planned or market economies, it is still the foundation of normal cost analysis. Modern industrial systems still do not normally include in the cost of what they produce such diseconomies of production and distribution as the spewing off of effluents into the air or the overloading of the land with solid waste or the lack of any charge for eventual disposal of the used-up goods. Thus they pass on a hidden and heavy cost to the community where it is either met by higher taxation and public spending or by the destruction of amenity.

The costs cannot be avoided. The citizen pays either as consumer or as taxpayer or as victim. The political and economic problems raised by this inexorable and unavoidable price spring from the fact that different citizens are involved in the problem in quite different degrees. The taxpayer may be out of reach of the major pollutions and have no direct incentive to clean them up. Yet poorer citizens can hardly welcome an increase in consumer prices for daily necessities even though they might be glad of cleaner air. The calculus of who shall pay for what improvement is *the* political issue at the core of any policy designed to deal with hidden subsidies and external diseconomies that underlie many of the present methods of satisfying economic needs.

When we turn to the nature of these diseconomies, any very precise delimitation of the issues presents some problems. The three broad areas of pollution which we must examine—air, water, and soil—make up, of course, the three main constituent elements of our planetary life. The atmosphere of airs and climates, the hydrosphere of rivers, lakes, and oceans, the lithosphere from which rock has crumbled away over the millennia to give us our thin and fragile envelope of soil are all inextricably interwoven in all the systems which support organic life. Without their continuous interaction through all the eons of our planet's existence, our little biosphere of living things would quite simply never have come into being. It is thus perfectly obvious that the interactions of air and soil and water continue to be inescapable, that when any of them are used in a destructive way, the dangerous impact may be reinforced by the very closeness of their association with other systems. As a result, we need much more sophisticated estimates of the degrees and varieties of environmental insult. We may find ourselves reducing pollution in one area only to find that the other life systems are endangered still more.

If waters are cleansed of impurities by sewage systems which then burn the final product off as noxious gases in the air, the environmental return on the cost of the treatment plant can be nil or may even be negative if the gas in the atmosphere is more pervasive than the waterborne wastes and can, by combining with a wider range of pollutants, produce more lethal damage. Similarly, there is little gain in cleansing the land of, say, the organic wastes of feedlots by sluicing them down into aquifers and streams if the rivers fill up with overfed bacteria and algae and lose their life-giving oxygen. In short, the problems of pollution, like every other serious environmental issue, demand multiple concern and multiple correction. It is, if you like, simply one more instance of the general proposition that technical man's interventions in the natural order are now on such a scale that "tunnel vision," the wholly specialized response, the single-thrust intervention all risk creating as much damage as they seek to repair.

Or take another issue—land use. Since modern man wants to use his fixed supplies of land for farming, for industry, for waste disposal, for roads and movement, for cities and settlement, for recreation, for contemplation—the risk grows that as his numbers increase, one good use will impinge on ten others. This makes the policies of urban, regional, and total land-use planning an inordinately convoluted affair. But so is nature. If man is to make use of his land, once again he cannot hope for single-shot, simple-minded solutions.

Nor is any policy for resource use, above all for energy, likely to come up with any more straightforward answers. Here, too, shortages, costs, price signals, alternative supplies and uses create a kind of ecosystem of their own of fluctuating interdependence. We cannot expect it to be otherwise, once we advance beyond the dangerous simplicities of past practice. On the other hand, although definition is more difficult, we can hope, in the longer run, for more effective solutions.

Of course, it is easy to get testy with the apparently limited and humdrum scale of many of these solutions. Mankind will not perish, we say to ourselves impatiently, if a little more biodegradable waste slips through the activated sludge of a secondary treatment plant. But the matter is not so simple. The total interaction of atmosphere, hydrosphere, and lithosphere in underpinning so fragile a system as our world of living things means that over time, even in detail, it is dangerous to cheat. Like the small stone wedged under a great rock which prevents the unleashing of an avalanche, our "details" may be precisely the catalysts of infinitely larger processes and risks. We may begin with humble coal and common sludge. We could end with the future of humanity.

Air Pollution

Air pollution may at this moment be going through a trough. Its health effects are less dramatic than they were a hundred years ago when there was more dirt and soot in the air, and fog swirled through the winter streets of industrial cities. If we want to get the authentic feel of what it was like to live under fog in London, then the world's largest city, there is no more convincingly murky picture than the opening of Charles Dickens's great novel *Bleak House*:

Fog everywhere. Fog up the river, where it flows among green aits and meadows; fog down the river, where it rolls defiled among the tiers of shipping and the waterside pollutions of a great (and dirty) city. Fog on the Essex marshes, fog on the Kentish heights. Fog lying out on the yards and hovering on the rigs of great ships. . . . Fog in the eyes and throats of ancient Greenwich pensioners, wheezing by the firesides.

This is the quintessence of fog. But it swirls through the urban novels of Balzac and Zola. It haunts Dostoevski's city streets. At the very end of the century, Sherlock Holmes still lived in Baker Street in an apparently unending succession of foggy Novembers.

But our relative improvement may not last—and this for two reasons, both of them concerned with the central fact in the consumption patterns of modern man—the fact of combustion. A very large part of his leaping demand for more energy is satisfied through the burning of fossil fuels, above all, coal. A very large part of his personal mobility and entertainment is tied up with the internal-combustion engine. In fact, we must almost think of him as a new species of centaur—half man, half automobile—and it is the heavy breathing of his motorized half that pollutes the air, invades the lungs, and builds up the smog in cities.

There are, of course, other forms of pollutants from the industrial sector. The chemical industry has vastly increased the variety and exotic nature of airborne effluents. Toxic substances like mercury or asbestos or lead, which used to be, in the main, poisoners of those working in particular industries—the "mad hatters," for instance, who breathed in mercury as they fixed the felt—are now spread more widely in the atmosphere by a much greater variety of uses and technologies. There are also a whole range of air pollutants in agriculture which will be examined on a later page. But the main source of industrial air pollution is combustion and the two chief causes are the generation of electricity and the motorcar. If we take the United States as the country using most energy per capita today and already demonstrating levels and types of energy use toward which, if nothing changes, other industrialized states may increase their own production and consumption, we find that as the seventies begin, pollutants in the air, coming in the overwhelming majority of cases from cars and power stations, amount to some 200 million tons a year, about a ton for each living and breathing American.

Electricity generation is, however, expected to treble by 1990. Coal will still be providing at least half of this vastly increased flow. Yet, even today, half the sulfur dioxide comes from electricity plants fired with fossil fuels, 50 per cent of the oxides of nitrogen, 25 per cent of the particulates—fly ash and soot, even a measure of radioactivity. If America had to endure a tripling of these effluents over the next few decades with unchanged technologies, air pollution would clearly become unacceptably worse.

These are the risks. Are there alternative technologies which may lessen them? One preliminary point should be made. Although power generation *is* a major pollutant, this is partly because it has taken to itself all the earlier pollutions produced by each industry working under its own steam with its own furnaces and by commerical premises and households heating and lighting themselves as in the days before power companies, central electricity boards, and high-tension grids. The concentration of pollutants at single power stations enormously simplifies the question of control and offers economies of scale entirely unprocurable among a multitude of separate, belching, polluting industrial plants each responsible for its own energy.

Generating plants powered by natural gas or petroleum with a low sulfur content do not produce much air pollution. But in spite of new finds and new methods of transporting natural gas, it is not a major resource. Petroleum with little sulfur in it is also scarce and purifying it at the refining stage adds at least 10 per cent to its cost. In America, gas and petroleum together provide only 35 per cent of the fuel for powerhouses. The major problem is with coal.

It is abundant enough, although to get it many countries will pay a terrible environmental price in scarred and abandoned landscapes. According to some estimates, even if the whole world's demand for energy went up by 5 per cent a year and were wholly satisfied by coal, there would still be enough to last for at least another century—by which time it is not, according to past technological experience, wholly irrational to expect that alternative technologies will have emerged. In fact, as we shall see, they already exist in the shape of various ways of producing nuclear power. But even under the most optimistic estimates, energy produced from other sources than fossil fuels—including all nuclear plants and hydroelectric stations—is not expected to amount to more than 12 per cent of world output in 1980 and still only a quarter in the year 2000.

Most of the good hydroelectric sites in developed countries are already in use. Other nonpolluting technologies—such as power from the tides or thermal or solar power—can be used and possibly expanded. But few people see them as available in a major way in the next decades. The problem of coal looks like being the chief pollutant problem in power and hence in the industrial sector until the end of the century.

But a lot of things can be done to make it cleaner. Some sulfides and particulates are treatable and removable at the mine itself. The injection of limestone may remove most of the sulfur dioxide from smokestacks. However, it can leave a problem of how to dispose of the resulting compound, calcium sulfate. It is also about 20 per cent more costly than the alternative process of building very tall chimneys and relying on the winds to disperse the sulfur dioxide without harm. This only adds 1 per cent to the fuel costs. But although the gases may be safely and easily removable in this way, they can fall in rain and snow on other places in the form of sulfuric acid.

Another hopeful technological advance which can also be introduced at the mine itself is the gassification of coal. This process, for which some governments have made money available for further research, has the advantage of producing the equivalent of sulfur-free natural gas, giving off no soot or fly ash and eliminating all carbon monoxide emissions. As restrictions upon the effluents produced by using coal are strengthened and the cost of all polluting fuels goes up, either as a result of fines or charges, the extra costs—whatever they may be—of processing the coal into pollution-free gas will be covered, and this new technology may well become the basis of much of the developed world's generating capacity as demand goes up in the coming decades.

The second most serious polluter of the air is, of course, the pride, joy, and workhorse of modern man, his motorcar. In the United States there is now virtually one automobile for every second citizen. Their output has been growing recently twice as rapidly as population and they account for at least 40 per cent of the nation's vast consumption of oil products. They are responsible for the largest output of oxidized carbon —the poisonous carbon monoxide, the ubiquitous carbon dioxide. They also produce the largest share of oxides of nitrogen as a consequence of the very high temperatures of combustion needed for ever more powerful cars.

Yet cars are clearly the most popular of all consumer durables (or energy slaves). They are containers as well as movers—for shopping, for children, for the whole family on holiday. They appear, in the wake of affluence, as man's first technological love. They will thus not easily be banished. Can anything be done to make them rather less lethal?

A number of things have been tried. Blow-by gases which escape from engine cylinders have been recycled since 1963. This process eliminated about 20 per cent of unburned hydrocarbons and carbon monoxide emissions. The new muffler to be introduced by 1975 at the latest may eliminate 90 per cent of the emissions from new cars, and since the catalysts are poisoned by lead, their use will eliminate the 2 million tons of lead now used annually as gasoline additives. Yet the mufflers' performance depends upon good maintenance. It is unlikely to be effective for more than 50 per cent of the.car's working life. If over the next decade, the number of cars in America doubles again, the amount of pollution looks like being much the same but from a higher number of cars.

It looks, therefore, as though other policies will be necessary. One concerns cost. The worst pressures from automobiles both in terms of air pollution, congestion, and disruption occur in cities and certainly no commuter pays the full economic cost of bringing his car—often a single driver in a station wagon—into the scarcest real estate in the world, the center city. A system of tolls or metering systems could lessen this particular pressure, especially if it were accompanied by subsidized and agreeable mass transit. At present, in many countries, the subsidies go all the other way—to the private motorist and the road building he requires.

If for intercity transport a switch could be made to electric trains, one calculation estimates that, for every 100,000 passenger miles, there would be a saving of 16,000 tons of gasoline, and net decreases of 8,000 tons of carbon monoxide, 1,600 tons of hydrocarbons and 320 tons of the particularly noxious oxides of nitrogen. In most developed countries, new technologies for rapid transit—French hover trains, the Japanese high-speed Tokyo-Osaka railway, San Francisco's Bay Area Rapid Transit system—are being considered or carried through. In fact, the motorized impasse in the cities is moving many governments toward a complete rethinking of their transport strategies.

Since, however, the love affair of man and automobile is clearly not over, there is urgent need for research into smaller, nonpollutive single cars. There are 100,000 electric vehicles in Britain. Diesel oil is being used. Natural gas is a possibility. Even the steam turbine has resurfaced.

What is clear is that the pollutive nature of the private car, combined with the steady shrinking in oil reserves, compels a rethinking of the technologies of private transport. People want to move. They want to be private. They want it so badly that they accept as a by-product, a fantastic accident rate—50,000 deaths a year in America—which if it came from typhus or cholera would cause a revolution. But whatever the desires of individual consumers, necessity is dictating a different type of car and research has to be geared to produce it.

Can we put a cost on anything so pervasive and elusive as the damage brought about by airborne pollutants? Once again, we turn to American figures. The reason is not that damage is worst in the United States but that Americans have made the most intensive attempts to give it a price tag.

One approach is to estimate what it would cost to clean up buildings, home furnishings, and clothes all marked by the steady rain of soot ash and corrosive acids. To this we must add vegetation since the blighting of crops and gardens around chemical industries and power stations is another "cost" unrepresented in the prices paid for power or plastics. Within this general calculus one can make fairly specific estimates on one particular set of values—those of real estate. In all societies where there is private or cooperative home ownership, the decline in property values can be used as an index of the economic cost of dirty air. Studies conducted in the United States in the mid-1960s covered eighty-five cities and made an attempt to correlate the local concentrations of sulfur dioxide, ash, and soot in the air with the decline in property values for private residences. The not altogether surprising result was to find a fairly clear relationship between faded paint, anemic shrubberies, bad smells, and dirty walls on the one hand and, on the other, the price a purchaser was prepared to pay either to buy or rent a house. The dirtier the air, the cheaper the building.

The study even attempted to arrive at precise marginal costs, reckoning that every 1 per cent increase in airborne dirt led to a decline in property value of 0.8 per cent. On this basis, it could be estimated that if one used a discount rate—in this case, 12 per cent—to assess what the property would have earned if it had not been polluted and used this rate as a multipler of the estimated damage, one could arrive at an annual loss in property value of about $620 million.

Using such techniques as these and extending its coverage to every form of quantifiable air-produced dirt, the United States Council for Environmental Quality has come up with an estimate of damage to vegetation and to the works of man of the order of $4.9 billion a year.

But this is not the end of the story. Air loaded with dirt and poison damages organisms much more delicate than public buildings, suburban drapes, or prized cherry trees. It enters human lungs and reduces human resistance to respiratory complaints. For instance, in the last lethal smog to descend on London in 1952, over three thousand people died of respiratory diseases.·

Admittedly, not all the costs of such diseases can be laid to polluted air. For many sufferers, the most personal of all air pollutants—the cigarette—has already done irreparable damage. But we encounter here, as in most forms of environmental damage, the phenomenon of the "threshold." Lungs can clear themselves of a surprising amount of debris. But little by little their ability to do so is cumulatively choked by the persistent intake of ash and soot and acid. At some degree of accumulation, the self-cleansing powers are exhausted.

The victims of chronic bronchitis or emphysema or lung cancer have two sets of costs to pay. There is on the one hand their loss of earnings. This turns up in society's actual, though not officially calculated, balance sheet as a loss of skilled output. On the other lies the whole range of medical costs which in personal terms may or may not be mitigated by forms of public and private health insurance but is, for society as a whole, a diversion of medical and hence expensive manpower and education to repair damage which need never have been done in the first place. The Council on Environmental Quality actually puts at $6 billion a year the economic costs entailed by lost days of work through illness and the mounting up of doctors' and hospitals' bills.

It is perhaps at this point that a general difficulty in environmental calculus should be examined. By our relative neglect of external diseconomies, by treating air and water as "free goods," we have accustomed ourselves to a method of measuring our wealth which gives us all the "goods" and tends to leave out all the "bads." Indeed, it includes such irrationalities as making no subtraction for days lost or lungs congested but includes as a "good" doctors' earnings for putting the troubles right. The figures proposed by the Council for Environmental Quality suggest, for all their inevitable imprecision, that America's Gross National Product should, on account of air pollution alone, be reduced by over $10 billion a year. But in formal accounting, such costs are invisible. As a result, citizens do not know what in fact they are paying and losing as a result of unclean, dangerous, and often stinking air.

The first need is therefore for society to recognize that there *is* a bill to pay and that it will either be met by the orderly diminution of air pollution or will continue to mount in the shape of lethal and unpredictable increases in the amount of damage from polluted air. Once the decision *is* taken, there is no shortage of antipollutive techniques. In the United Kingdom the Clean Air Act of 1956 forbade a large range of uses of coal unless its pollutants had been reduced and established smoke-free zones where no untreated coal could be used at all. The amounts of smoke and sulfur dioxide have since steadily gone down—although population has risen by 10 percent and energy consumption by 17 per cent. For the sun-starved British, a particular bonus in central London is a 50 per cent increase in the amount of winter sunshine.

Another line of approach is to establish effluent standards—the amount of particulates, sulfur dioxide, oxides of nitrogen, and so forth —that can be tolerated and to impose Draconian fines if the level is exceeded. Provided an honest and efficient inspectorate can be guaranteed and the penalties are higher than the costs of remedial action, these controls can work and have been one element in the decline of industrial air pollution in Britain. Alternatively, industrial establishments and power stations may be compelled to pay a charge, an "effluent charge," corresponding to the amount of dirt they put into the air—the higher the emission, the greater the costs. This procedure has the advantage of hastening research into nonpollutive technologies. If the atmosphere, which has hitherto been a "free good," in other words, a free sewer for industrial dirt, now has a cost attached to it, the engineers designing new factories and power plants must take this cost into account and have, for the first time, an inducement to minimize its use. If in the process of *not* putting the old dirt into the sky, they find that some of it can be actually reused and resold, the new technology may even be more profitable than the original mess.

Bans, legislation, fines, and effluent charges seem likely to clean up a large part of the pollutions. But there may still be need for some forms of public expenditure. Here we encounter the problem that citizens, too, are accustomed to the idea of air as a "free good," are not used to the idea of paying to keep it clean, and can even resent and resist the idea that public expenditure may be needed for atmospheric spring cleaning. This resentment is particularly acute when it is proposed to use taxation as one means of securing the needed funds. Since citizens with higher incomes have usually already removed themselves to cleaner, healthier atmospheres, they are likely to be especially unable to see the connection. This, however, is a general problem underlying all systems of progressive taxation which, as a redistribution of income from the fortunate to the unfortunate—in this case from the clean to the dirty—arouses less than wholehearted enthusiasm among those who foot the bill. The cleanup is not *our* cleanup. Why should we be concerned?

There are two points to be made here. The first is one of simple equity. Effluent fines and charges will inevitably be passed on in some consumer prices, and poorer members of society will pay more, relatively speaking, than the rich if electricity prices increase in the wake of cleaner air. If some of the costs were covered out of public tax-raised expenditure —for instance, by putting larger public funds into research for antipollutive technologies—the result in market economies would simply redress a balance likely to be weighted against the less wealthy.

The second point is a broader one. There can be no civilized society —market, mixed, planned—unless the more fortunate and better-paid members of that society make a proportionate contribution to the broad social goods represented by services and amenities the individual citizen cannot provide for himself. The time has now come for society to recognize that the needs of a healthy environment—clean air, clean water, unpolluted soil, decent cities—are public, social goods on a par with good order and good education and must be provided because communities cannot survive without them. The citizen has to learn to see their provision as a primal need which nature may have provided free in the past but which now enters man's essential budget of civilization. It was right that, as after the London smog of 1952, the decision should have been taken that clean air and clean water are part of the essential needs of decent existence, and the cost of providing them must be met. This is a community decision, a political decision—like the nations' general commitment to education or, less fortunately, to defense. But once it is taken, there is little doubt that the task of cleaning up the mess—or of preventing it in the first place—will be found to be manageable. Indeed, twenty years from now citizens may look back with wonder and ask themselves why so valuable a social effort was taken with such unbelievable delay.

Pesticide Control

About 80 per cent of the airborne pollutants in developed countries occurs in the course of combustion—external combustion in industry and power, internal combustion in motorcars, even more "internal" combustion in what may well be the most lethal of all breathable pollutants—tobacco. But we cannot disregard the quota provided by processes on the land.

In the years since Rachel Carson's epoch-making book *Silent Spring* the dangers she exposed have been analyzed, researched in ever greater complexity, in part modified, in great measure accepted. The reason, of course, is that Rachel Carson directed most of her warnings not against pesticides as such but against the great sloshing wave of overuse which followed on the first intoxicating discovery of what some chlorinated hydrocarbons—of which DDT is the best known—could apparently do. They first appeared capable of wiping out the multitude of pests which beleaguer mankind—some of them, like fleas, carrying plague; some, like the mosquito, carrying malaria and yellow fever; others, like the screwworm fly, dangerous to useful domestic animals. The new poisons seemed an unmitigated good and if they were so effective even in small quantities, might they not be even more so if the dose were increased? This was the euphoria behind such runaway programs as saturation spraying from the air with poisons whose effect was still barely understood.

Again and again in Rachel Carson's book the plea was for more knowledge, more research, more exact information about performance and consequences. In the wake of her plea, and partly as a result of it, far more research has been undertaken. What has emerged is a picture of great complexity which vindicates her moving indictment of ignorance and overuse.

We find, in fact, all those elements of complexity in natural systems which first underpinned the emergence of organic life and now sustain it. We find once again the phenomenon of the threshold. Few people suspected in the 1950s that substances such as DDT had the same capacity as such well-recognized metallic poisons as mercury to become steadily more concentrated as they move up each stage of the food chain. Fish eaten by birds can thus be passing on concentrations which, by hidden degrees, have already passed beyond the birds' tolerance to DDT. The poison is further concentrated in their organisms and either they cease to reproduce or lay eggs too thinly shelled for hatching.

Then we find the bewildering degree of interdependence of natural systems. Toxic materials cannot be put into either air, water, or soil without their reappearing in the other two. This interdependence extends to the creatures living in the three elements. At one time it seemed reasonable to souse large areas in the southern parts of the United States with a lethal mixture of dieldrin and heptachlor, two of the most poisonous chlorinated hydrocarbons, in order to eradicate the fire ant. The poisons applied at a level of two pounds per acre were literally allowed to fall where they might. The fire ant survived. A very wide variety of other species were decimated. It was at this point that the question began to be more widely raised whether, by indiscriminate killing, another of nature's delicate balances might be destroyed—the balance between pests and the other creatures in nature which prey upon them. Suddenly eradicate one predator from the food chain and it may be found that a population explosion of another pest occurs at another point along the chain.

The third aspect of complexity is nature's inexhaustible variety—of soil, of climate, of species, of plants. This is indeed the glory of man's natural environment, and one of the dangers of monotonous and undifferentiated urbanization is the tendency to standardization and uniformity, which can be a cultural blight on the human psyche. But in nature standardization, in the sense of uniform treatment for uniform conditions, can be more than a blight. It can be literally a killer. Doses of weed killers or pesticides which are correctly balanced for one kind of air condition, climate, soil, and plant ecology can be lethal in neighboring conditions or even in the same conditions on a more or less windy day. It is perhaps here, more clearly than anywhere else, that we see the dangers of any oversimplified acceptance of science's essential techniques of abstraction and generalization. In a range of inorganic and man-made things, laws of measurement, stress, balance, force, and counterforce all hold good. It is perfectly possible to produce valuable and reliable standardized products according to specification—from structural steel to prestressed concrete to homely bricks and plywood. But one of the essential points about agriculture is that much of its fertility, range, and resilience depends, precisely, upon its lack of uniformity. Unless the effects of attempted standardization, for instance in monocultures, are purposefully offset by man-induced variety, fertility can be endangered. As we shall see when we look at the problems of tropical agriculture in developing lands, it is precisely our lack of accurate, scientific information about so many of the basic facts—of soil, of plant physiology, of climate, of water supplies, and of all their interactions—that makes so hazardous the sudden introduction of the products of high technology and science. Research has been, in the main, concentrated on the agricultural problems of temperate areas or on the few agricultural products designed for export to developed lands. There simply is not enough accurate knowledge of local conditions to devise the needed variety of controls, monitoring systems, fallback positions, and workable alternatives that can prevent a large thrust of intervention—be it pesticide, herbicide, or even improved strains of plants—from risking unacceptable damage. Fortunately, as we shall see, policies for purposeful variety *are* available and are likely to be more widely used as ecological understanding spreads among farmers.

If we try, in our approach to pesticides, herbicides, and, indeed, fertilizer use as well, to bear these three principles in mind—threshold, interdependence, and complexity—it is clear that the impact of the new scientific means of controlling nuisances and increasing productivity varies tremendously according to their durability and toxic effect.

Some pesticides are short-lived but are highly toxic; for example, malathion and parathion last only a month or two but are quite indiscriminate killers. Most herbicides also last for relatively short periods, from a couple of months to a year. Other pesticides may not have a high level of primary toxicity, but last so long that they have cumulative effects, and as a result generate a whole variety of delayed consequences difficult to predict. This is the problem with the chlorinated hydrocarbons, all of which—DDT and, more lethally, dieldrin and endrin—accumulate along the food chains in increasingly concentrated forms. Furthermore, the persistence of this class of pesticides allows insects enough time to develop immunity to them and even to acquire a positive affinity for the more widely used hydrocarbons.

Finally, the pesticides can be highly toxic and virtually permanent.

These are the nondegradable ones—for instance, all the various compounds of mercury, arsenic, and lead.

Can we, out of all this variety, establish some pattern for use—or nonuse? There seems to be no case at all for the nondegradable poisons. For instance, seeds coated with mercury to prevent fungus attacks may enter the food chain in a variety of ways; in at least one case such treated grain was accidentally fed to hogs with gruesome results for the human beings who ate their meat. Many countries have now banned the use of this mercury coating.

Should less toxic but still enduring chemicals also be banned? In the United States, Michigan has banned DDT. Even more significantly California, which, with only one-fifteenth of America's agricultural land, but using one-third of its pesticides, has introduced major restrictions. The Swedes have imposed an official "pause" in its use for two years to allow for a full scientific study of its consequences. Canada's strict policy of licensing has reduced use by about 90 per cent. Some countries prefer voluntary control systems. But corporations and individuals complying with the controls may then find their care and costs undermined by less scrupulous competitors.

But even if we simply proceed with the banning of a whole range of existing persistent pesticides, the problem is not necessarily solved. There *are* pests. They *do* eat the crops. In parts of Asia, up to a quarter of the harvest may go to feed not humans but predators. Corn output may increase by 60 per cent, potatoes by 70 per cent if their resident pests are eliminated. How do we avoid being as "tunnel minded" in our retreats from technology as we have too often been in our sorties and excursions?

There are some alternatives. Productivity began to spring up in the United States in the 1930s before hydrocarbons had left the formula stage when some short-lived and naturally produced insecticides, such as pyrethrum, were brought into use. Their disadvantages are their effective life of only a few hours and the requirement of repeated spraying.

There are alternatives to chemicals in the shape of biological forms of control. True, they have to be used with even more care than the chemicals. A mutant strain, running out of control, could be a more formidable danger than the chemical elimination of a single species. Yet, paradoxically, they may also offer the chance of avoiding damage to more than one species. In other words, they may turn out to be much more specific than organic synthetic pesticides. The screwworm fly which infects cattle has, for instance, been controlled by irradiating male flies and rendering them sterile.

Another possibility is the introduction of a specific predator to attack a particular plant or animal pest. The Australians brought a spreading ruin of cactus under control by introducing the cactoblastus, a South American moth whose larvae killed off the cactus. There is risk in this, of course. A predator or virus brought in to cure one evil may set off a dozen others. The more specific the relationship between predator and prey, the safer the experiment.

Another approach to biological control lies in carefully learning the growing and maturing patterns of both plants and insects. Then crops can be chosen and planted in such a way that they can be harvested before the expected onslaught of local insects or microbes.

So far, the most effective type of biological control, based upon a close and understanding cooperation with nature's own forces, has been the development of strains resistant to local pests and diseases. In this type of research the airs and winds, the cycles of vegetation, the rainfall, the soil types, the natural predators and their place in the local food chains —all have to be fully understood if resistant strains are to be successfully developed. Techniques of pest control, whether chemical or biological, can be safely used only within the context of such knowledge. Aerial spraying over vast acreages with lasting lethal poisons lies at exactly the opposite pole from this careful, watchful, respectful approach to nature's own complexity and from the attempt to devise curative measures that match.

This close and careful adaptation of means to ends is part of what we should mean when we talk of scientific farming. Agricultural science calls for a rigorously researched approach to the realities and interactions of air and climate, of water and soil and plant. The answers will not be broad generalizations. They are more likely to be "minute particulars"—specific instructions about what can or cannot be done in particular places at particular times with particular things. And as these particular control techniques are discovered and introduced, there is a strong case for a balanced phasing out of the general use of such long-lived hydrocarbons as DDT, dieldrin, or endrin.

But the need to establish what these particulars are and the overwhelming need not to phase out all present forms of defense before new ones are available underlines the plea made by Rachel Carson for more knowledge and more research. Probably much less than 1 per cent of the developed world's research money is being devoted to work on the alternatives to the massive use of long-lasting poisons which still dominates so much of so-called scientific, modernized farming. If deadlines were set for the beginning of a phasing-out program for persistent pesticides, clearly the stimulus to necessary research would be enormously increased.

With this need goes the necessity for training far more men and women to enter the field of truly scientific agriculture. As farming becomes more complex, the training of more agricultural scientists and of better-educated farmers and extension workers becomes not a luxury but a driving need. Traditional farming methods were not unscientific. Indeed, they were based upon one of science's most powerful tools— experimentation, which, in this case, is simply called experience. But there is a limit to productivity by traditional farming; for instance, the traditional Indian farmer's output per acre is a hundred times lower than that of the Japanese. Just as the neolithic farmer very greatly increased the soil's productivity by moving from the gathering of wild grain to the growing of cultivated seed, so today, the scientific revolution is making possible another leap upward in output. While the achievement cannot be doubted, it depends upon profound interventions by man. Hence farmers must know much more about the effects of their new technologies.

We shall return to this necessity in the context of farming in developing lands. At this point it remains only to point out that the economic implications of cleaning the air in the wake of farm pollutants are not much different from those incurred in cleaning up industrial air pollution. More public money is needed for research. There must be more reliance on a variety of solutions and on more careful calculations of the balance between the continuance of some pollution and the degree of strain the environment can carry without damage.

If a greater variety of pest-control methods and a more careful estimate of a balanced outcome make farming more expensive, the consumer will foot the bill. However, it is possible and indeed likely that with much sounder husbandry and more lasting productivity of soils the consumer will ultimately stand to gain from more reliable food supplies not threatened by sudden catastrophic, chemically induced disasters. If immediate food costs rise inordinately in the short run, the poorer members of society will need larger protection—but this issue of redistribution and social justice is raised by every issue of environmental improvement.

Water Pollution

It is a paradox that man has tended from earliest times to dispose of his wastes in the water courses from which much of his drinking water is to come. But under natural conditions, rivers have very considerable powers of self-cleansing. The flow of water scours the detritus—of salt and soil and sticks and stone—out to the oceans. Bacteria use the oxygen dissolved in the water to decompose organic wastes and in turn are consumed by fish and water plants, who return oxygen and carbon to the biosphere. The only real risk under these simple conditions is that some of the minute bacteria will get into someone's drinking water and give him one of the very large range of intestinal diseases which over the millennia have been a major human scourge. This remains the principal pollution in most of the world and is increasing with rising population.

But as mankind moves on from the Arcadian simplicities of farming and colic and enters the new urban-industrial order, the problem of waterborne wastes becomes much more complicated. First of all, industry brings thousands upon thousands of people together in urban concentrations. "Natural" systems of sewage disposal down the rivers become grossly overloaded. Then industrial processes can very greatly increase the range of materials which bacteria cannot deal with (the nonbiodegradable kind)—and some of them are poisons, particularly compounds like cyanides or minerals like mercury and lead. These, piled in industrial tips over the landscape, may also, by seepage, release their poisons into underground waters or neighboring streams.

Then again, even the organic (or biodegradable) wastes—from municipal sewage, from pulp and papermaking, from feedlots—can overload the river's available supplies of dissolved oxygen. The bacteria use it all up as they decompose the gobs of sewage. Oxygen levels fall. Sometimes there is simply none left, and since all aquatic life requires oxygen, the river loses its capacity to carry living things and may flow on for miles as a dead and stinking sewer. The slower the river's flow, the greater is the risk. All the rivers round Japan's papermaking city, Fuiji, are in this condition.

Even when some oxygen does remain after the bacteria have done their job of decomposition, there is a further danger. Decomposition reduces biodegradable sewage into simple molecules containing basic elements—potassium, phosphorus, nitrogen—and other nutrients. Add them to water and aquatic life receives the equivalent of a large dose of fertilizer. In many rivers, explosions or "blooms" of bacteria and algae eventually deplete the remaining supplies of oxygen. Then, with oxygen gone, other bacteria which do not need it—the anaerobic ones—go to work on the remaining wastes, giving off stinking gases such as hydrogen sulfide.

Bodies of water with little flow and a lower capacity for recovering oxygen are particularly exposed to this process, known as eutrophication. The natural washing of silt and nutrients into lakes tends to make them more shallow and to change the kind of life they can carry. But modern siltage and effluents enormously speed up the process. Lake Erie is the most notorious example but many lakes in Europe are dangerously short of dissolved oxygen and even landlocked seas can be affected. The oxygen content of the Baltic measured at Landsort Deep has fallen by 250 per cent since 1900 and today oxygen is virtually exhausted in this stretch of the sea.

Furthermore, wholly new pollutants, invented by man and never found in nature, are getting into the waterways. Pesticides are only the most notorious of a vast multitude of man-made compounds which by now number at least half a million, with new ones arriving at a rate of five hundred a year. Since it is costly to test the potential toxicity of new substances, the chance of a thorough investigation is, at the moment, not high. Man's water supplies are constantly being contaminated with substances, the long-term effects of which are largely unknown.

And if they include substances which either become concentrated in the organism over time or, by repeated ingestion, weaken essential biological defenses, then we reach once again the threshold of tolerance beyond which recovery itself is in doubt. In every industrial country, there is immediate need for a far more rigorous registration, testing, and inspection policy for this sector of industrial activity. The ultimate folly in the pursuit of so-called scientific technology is to leave out science's central claim and—on occasion—achievement and that is to help man to know what he is actually about.

Last of all, the steady rise in demand for energy means, in some countries, a very large increase in thermal pollution. Water used as a

coolant in generating stations—and in some other industrial processes—
is poured back into the rivers, where by raising the temperature it speeds
up certain biological processes and imposes sharp changes on aquatic
life. Since the feeding and breeding of all types of fish are affected by
temperature, a raising or lowering of the heat of the water has profoundly
disturbing effects, eliminating some breeds, overstimulating others, and
in some conditions, destroying the lot.

What is the scale of all these risks to man's use of his life-giving
waters? The developed world's defenses against waterborne diseases
seem to hold. In spite of a steady increase in population and of the
number, though not the density, of people gathered in urban settlements,
the nineteenth-century sanitary revolution—with its reservoir building,
its control of catchment areas, its primary treatment plants, and its
chlorination—still underpins the health of modern urban man. In spite
of very occasional outbreaks in overcrowded holiday resorts, cholera and
typhoid have vanished from the developed world. Typhoid is a seven-day
wonder. Gastroenteritis in infants is no longer a dreaded killer. Since in
countries like Britain at least a third of all publicly supplied water comes
from rivers that are also channels for wastes—in the United States the
proportion is even higher—it is clear that the continuance of reliable
drinking water is a considerable technical achievement.

But, if we are holding the line on disease, the problems of all the
various wastes—solid, degradable, thermal, radioactive—have been
made much more formidable by the recent twenty-year-long industrial
boom, by the explosive growth in the use of energy, and by a vast increase
in the scale of the petroleum and chemical industries.

One way to measure waterborne effluents is by the amount of oxygen
needed to decompose them—the so-called BOD or biochemical oxygen
demand rate. Taking this as a measurement, effluents from industry in
the United States seem to be using up three times as much oxygen in the
water as the sum of all the effluents from municipalities with reasonable
sewage systems. Moreover, some 40 per cent of the water treated in
municipal sewage plants comes from industry and it has been estimated
that over a quarter of a million of these industrial plants produce effluents with which municipal sewage systems are not equipped to deal. If
one includes in industrial pollution the 430 billion tons of heated water
released every year into the estuaries, lakes, and rivers and adds the
effluent from the concentrated industrialized production of meat and

poultry in feedlots—which is said to be the equivalent of wastes from 800 million people—industry emerges as the main source of waterborne effluents, with municipalities supplying most of the rest.

Should agriculture be added as a major source? The toxicity of some of its by-products is not in doubt. Between 1960 and 1963, ten to fifteen million fish appear to have been killed by the pesticide endrin leaking into the Mississippi River. In 1969, two hundred pounds of endosulphan, falling off a Rhine barge, took a tremendous toll of fish, the dead bodies floating across international boundaries to horrify the downstream Dutch. In some areas industrialized production of meat and poultry in feedlots adds enormous wastes to the waterways. But there is the further question whether one of the nutrients most responsible for explosions of bacteria and algae and hence for the risk of oxygen depletion may not in fact be nitrogen running off the lands of overenthusiastic farmers, pouring on artificial fertilizers to make up for the loss of natural nutrients in the soil.

The experience of Britain may be relevant. It has one of the most highly fertilized farming systems in the world. Only Japan and the Netherlands use artificial fertilizers more intensively. Every year British farmers apply at least three times more nitrogen per acre than the Americans—the Dutch use ten times more. If the run-off of nitrogen was universally a serious factor, one would expect a significant showing in British waterways. But the recently appointed Royal Commission on Environmental Pollution concluded in its first report that, in conditions of good husbandry, run-off is not a major problem. One should add, however, that the British example, like the Dutch, has some special elements. There are no large lakes. Rivers are short. The problems are different for enclosed waters and in long and often slow-moving rivers.

The British report is also not unduly alarmed by another risk of nitrogen run-off—the accumulation of nitrate in drinking water. When consumed by humans, nitrate can become the toxic nitrite, which reduces the blood's ability to convey oxygen—a condition to which babies are particularly susceptible. No major river system in Britain has anything like the one hundred parts per million which the World Health Organization has set as the threshold beyond which the risk of severe damage may begin. However, there is one area in which nitrate levels are so high that bottle-fed babies are provided with special water. Thus there *is* a problem, given Britain's very high use of nitrogen fertilizer but, apparently, it is not one of major proportions. Nor is it universal.

Should one conclude that America, using two-thirds less nitrogen per acre, faces more serious risks? One can only report a considerable difference of scientific opinion; in some areas public health authorities are much stricter than WHO and allow only ten parts per million for nitrates in the water. A number of wells and boreholes have been known to approach this level in Illinois and California. On the other hand, the International Joint Commission responsible for the Great Lakes has taken the view that the chief cause of algal bloom is not nitrogen run-off but phosphorus from municipal sewage. And this judgment is borne out by the experience of other countries in which, in spite of very high use of nitrogen fertilizers, municipal sewage systems are the major source of polluting nutrients.

Are there technologies available for dealing with this wide range of pollutants? Obviously the sanitary techniques of the past still work to produce drinkable water—even if it does taste of chlorine. Otherwise industrial man would suffer more than he does from waterborne intestinal disorders. Yet, as his numbers and his uses rise, the amount and variety of pollutants that he releases into the waters must make it increasingly difficult and expensive to secure the minimal amount of safe, separate drinking water. As Britain's Royal Commission points out, rising needs in Britain will compel the country in the future to make more, not less, use of its rivers in providing water for drinking or domestic purposes. Yet if they are growing steadily more dirty, their usefulness must decline. The capacity of a river to dilute pollutants and render them harmless is fixed by its size, so that if in the future a river is to receive twice as much effluent as it does now and yet not become less pure, the effluent will have to be twice as clean as present standards require. If the river is to be made still cleaner—as will certainly be necessary in many cases if it is still to be used to supply drinking water—an even higher standard of effluent purity must be achieved and this will require more thorough treatment at higher cost.

In the cheapest primary treatment of waterborne wastes, grit is screened out, scum removed, and other wastes allowed to fall in the sedimentation tanks to form sludge. The difficulty is that this simple process leaves at least 50 per cent of potentially oxygen-consuming wastes in the water. A lot of sludge is also still there at the end of the process and has to be added to the disposal problems already created by other forms of solid waste. Burn it and you pollute the air. Use it for landfill and you can run out of space or slip dangerous effluents to underground waters. The whole process is another simple reminder that we do not "consume" matter. We use it, discard it, and still have to cope with its residues and effects.

The next stage in technique—and expense—is secondary treatment. With the help of oxygen and bacteria in the sludge tanks, this treatment destroys most of the organic wastes that otherwise suck up the water's oxygen and lead on to algal blooms and bacterial explosions. But whether a sludge is activated in this way or just left as common sludge, it still has to be disposed of. However, there may be more satisfactory methods than simply dumping or burning. Chicago, for instance, produces 1000 tons of sludge a day and in 1970 it cost $60 a ton to put it into deep holes near treatment plants. But half this sludge is now dried and shipped to Florida where it is sold for $12 a ton to citrus growers. The procedure costs less than hole-filling and puts back into the soil some of the nutrients that have been washed out of it. There are also possibilities of sending sludge by pipeline to central Illinois to cover the hideously scarred lands left by strip mining and to fertilize marginal farmland. The cost of the operation —estimated at $20 a ton—could both restore nature and even cover some of the costs, an ideal balance not always to be realized in environmental affairs. However, in all such plans, toxic industrial wastes must first be securely removed. Otherwise, reuse simply spreads the poisons about.

Secondary treatment has, however, the disadvantage of leaving a lot of the more complex chemical pollutants in the water and also of doing little to remove the growing amounts of nitrogen and phosphorus. Once passed through secondary treatment plants, they tend to reappear in forms which are actually more appetizing and absorbable for algae and bacteria. This last trouble could be enough to wipe out most of the benefits of the whole costly system. It would be a tragic and disillusioning waste if, for instance, over the next five years the United States invested some $13 billion to give secondary sewage treatment to 90 per cent of its municipalities only to find, at the end of it, more algal bloom, more bacteria, and less oxygen in the water—precisely the evils which secondary treatment was designed to prevent.

In wealthy countries, therefore, there is bound to be a rising interest in tertiary treatment which aims at removing more than 95 per cent of all pollutants and leaving the water not far from drinking standards.

There *are* such technologies although many of them are still experimental. In the United States at California's Lake Tahoe, a lake of exceptional depth and color—but one threatened by the sewage of an eager tourist industry—a plant takes in 7½ million tons of lake water a day and after removing the ordinary pollutants and sludges goes on to use lime in the removal of phosphorus and to "blow" out the nitrogen (which usually appears in sewage in the form of ammonia) in a so-called stripping tower. The water then passes first through further separation beds to remove the last of the phosphorus and then at the end over activated carbon, which, with its convenient bonding points, combines with most of any remaining chemicals. But, as it can well be imagined, the costs per thousand gallons of effluent in the process are up to 25–30 per cent higher than in secondary treatment.

Which brings us to the central problem—how much can, should, or does a society want to pay for clearing up what proportion of its water-borne mess? We are, after all, accustomed to using water at very different levels of purity. A surgeon washing his hands before an operation insists on water that is purer than drinking water. It is convenient but not absolutely necessary to have bath water clean enough to drink. Few people mind a few impurities as they plunge into rivers and oceans for a brisk swim—although they draw the line when, as on Rome's summer beaches in 1971, the likely result is hepatitis.

Also, different groups need different standards. Some industries can recycle their own internal water needs very cheaply while releasing the untreated balance to the water courses. Fishermen probably need less purity than swimmers, duck shooters perhaps less still. We are once again confronted with the astonishing variety of human needs and the disturbing way in which those who cause nuisances are seldom the people who suffer from them. As a means of keeping good behavior entirely disinterested, this randomness is, no doubt, a key to the planet's ethical survival. As a principle in economics, it is decidedly awkward.

What, against this background of variety and contradiction, can we say about costs? The first point obviously is that no sane society can afford to discharge direct poisons into water systems. Take the case of mercury. The bacteria in silt and decaying matter convert it to methyl mercury, which becomes more and more concentrated as it moves up the food chain—from bacteria to small water life to big fish to man. In 1953 fishermen in the large, nearly landlocked bay of Minamata in Japan began displaying the symptoms of the Mad Hatter—diffidence, anxiety, irritability, hallucinations—followed for many of them by mental derangement and death.

The reason was simple. Shellfish in the bay had consumed methyl mercury, the fishermen had eaten the shellfish, and the poison, still more concentrated, had lodged in their brains. Since then, fishing has ceased in Minamata Bay and the lesson is obvious. Governments have to consider whether a range of poisons—acids, heavy metals, toxic compounds, a list on which lead and mercury figure prominently—should not be banned as effluents either to water courses or to dumps. Such a decision could expose industries allowing effluents to escape to the risk of heavy fines and thus encourage them to rethink their industrial practices.

There is also widespread discussion of the case for extending bans to such notorious polluters as the modern detergents. It is true that in the last decade they have at least been made biodegradable. In some areas, too, coastal waters are deficient in phosphates and sewage disposal could enrich them usefully. But in general the phosphates in detergents, especially near freshwater lakes, are still the major cause of algal bloom. As with industrial poisons, prohibitions on use would place on industry the task of finding nonpollutive alternatives. Some counties and municipalities have introduced such bans but satisfactory alternative products are not yet in sight. Since detergents have proved extremely convenient for the housewife and large switches in technology take time and research, the argument has been made that in areas where phosphates do real damage they should be phased out and public incentives be given to further research. ·

But beyond the field of poisons and so major a pollutant as the detergents, calculations of cost and benefit become more difficult. Highly industrialized states can survive for long periods of time with no more control than primary treatment for wastes and protection of drinking water. Japan has contrived to do so—although with less than universally acceptable results. Meanwhile, bathing may gradually decline. Little boys who could be swimming may be releasing fire hydrants or splashing in fountains or out skinning cats. The rivers along which lovers might formerly have wandered may be deserted as hydrogen sulfide replaces the old smells of springtime. But it is a bold economic calculus that balances the increase in sewage costs against the decline in lyric poetry.

We come to something more tangible when lakes and rivers are simply unusable for outdoor recreation. An economic model was recently constructed for the Delaware River Valley in the United States, balancing the cost of securing given levels of dissolved oxygen in the water against the income likely to be earned from increased recreation. The estimates, completed in the mid-sixties, suggested that a charge of only $2.55 a day for boating—over the next three decades—would cover the entire cost of producing a river agreeable enough for people to want to boat on.

But the rising pressure of both people and of effluents in wealthy lands over the next thirty years suggests that even the cleanliness that permits boating may not be enough. All water could be on the way to becoming so precious that only treatment at Lake Tahoe standards will be sufficient to keep people in health, let alone in happiness. Indeed, there may be a progression of "development" in which, according to numbers, urbanization, and degrees of industrialization—in both manufacturing and agriculture—a nation goes inexorably from primary to secondary to tertiary treatment, the costs expanding inexorably all the time.

But at some point along the progression some innocent economist or technician or even politician is bound to ask: Would it not be better not to put the effluents in the water in the first place?

This question suggests an interesting and useful standard by which to judge alternative policies for dealing with waterborne pollution. One can ask not which process cleans up the water most effectively but which makes it less likely that the waters will need to be cleaned up. It can be argued that of the two main approaches—spending to encourage better treatment or charges on the effluents that are discharged—the latter has the more radical effect. No doubt both approaches are needed. But innovative research and public policy may achieve better results on the preventive side.

There are a number of difficulties about the approach through incentives and subsidies given specifically for sewage treatment. Tax incentives, available only for treatment plants, have little impact on basic technology and, as we have seen, even secondary treatment is ceasing to be really adequate. A "more of the same" policy in both industrial and public schemes may only serve to fix treatment techniques which are beginning to be too expensive and also out of date.

Besides, they take little account of the variety required in effective policies of water management. Rivers, being organic and natural and part of the infinite variety of the planet's untamed existence, are not uniform bodies of water. Temperature levels change. Beds along the river differ in composition. Flows and evaporations fluctuate. An indiscriminate factory-based or municipality-based set of different treatment plants may be very far from an optimum—which is essentially a flexible—treatment of a whole river system. It is significant that in one of the first really effective experiments in the control of water pollution—the association or *Genossenschaft* set up in the Emscher Valley of the Ruhr early this century—it .was precisely the failure of two hundred different municipalities with their separate treatment plants that persuaded the authorities, after, as usual, an unpleasant epidemic outbreak, to accept the need for general management of the whole river.

This is the direction in which, it can be argued, any serious effort of water management has to move. There are two sides to the policy. Tbe first is to accept the principle that those who produce the pollutions must pay for the cost of cleaning them up. In the Ruhr system of *Genossenschaften,* which link the Ruhr and its tributaries into two interrelated, cooperating groups covering all five rivers, the charge is based on the effect of the effluent on the river's need for dissolved oxygen—its so-called biochemical oxygen demand (BOD)—and the threshold for discharges is fixed by the level at which fish can continue to live. Effluents are monitored and tested and according to the strain they put on the river's oxygen, the firm's charges are fixed.

This process has two advantages. It gives the *Genossenschaften* a steady income with which to maintain the least costly and most flexible instruments of water management. These include large integrated treatment plants, dams, and storage lakes from which water can be released when the rivers are low, the aeration of some sections, extensive land drainage, and even the use of the Emscher as a general sewage disposal canal with no more treatment than is needed to avoid damage to recreation or aesthetic enjoyment. The result of this integrated approach is a remarkably low level of costs. Parts of the system make money. For instance, the waterworks provide drinking water and make a profit although their prices are the lowest in Germany. Overall, about $60 million a year is spent by the *Genossenschaften* and this for a river system whose natural advantages—topography, amount of flow, regularity of water levels—are by no means considerable in comparison with other less well-managed rivers where costs are higher, with much poorer results.

At the same time, effluent charges encourage each individual plant to plan its own most efficient mix of treatment and discharge and it allows for a more flexible response to the river's own variations. Yet this is not, as some critics maintain, a "license to pollute." The fact of having to pay for effluents means a very steady pressure upon industry to invent nonpolluting and hence tax-saving technologies. And the interesting point is that when the techniques have been worked out, it is often found that the metals and minerals no longer wasting into the river are valuable by-products capable of being economically recycled. By getting away from the idea of the river as a free sewer and the material as "consumed," not simply used, the industrial process begins to look a little more like nature's own rhythms of using, degrading, storing, and reusing its constituent materials. The throwaway economy begins to give ground a little to the economy of conservation.

In Germany, for instance, usable sulfuric acid is recovered from the pickling liquor in steel manufacturing. The steel industry also recovers more from its discharges by putting them not into the river but into sedimentation ponds. Canning industries have found they can recover salable vinegar from what once were wastes. Paper industries, by shifting from a sulfite to a sulfate process, have reused their chemicals to such good effect that in modern plants, effluents can be cut down by as much as 90 per cent. One company has developed a process which converts its waste black liquor into activated carbon. This is then used to filter the paper mill's liquid effluent, which can then be reused in the mill.

Similarly, thermal pollution of water can, with careful overall management, be converted to some good uses. Ponds into which hot water is pumped for cooling have to be large but both the Russians and the Japanese are using them successfully for recreation and for fattening up fish like eels or carp which respond well to warmth. American power companies situated at the seashore are experimenting with oyster beds. Hot water used in irrigation can speed up germination. The International Joint Commission for the Great Lakes even believes that by heating the lower strata of Lake Erie and Lake Ontario, the water could be forced to the surface and assist in cleaning up the pollution. In all these cases, the move is from the concept of waste to the search for reuse. As such, it lessens the long-term disasters inherent in the throwaway economy and begins to respect and reflect, even if still only in "minute particulars," the finally closed nature of our limited biosphere.

There are two final points to be made about the need for the integrated treatment of rivers and lakes. The first is that it is very difficult to establish agreed standards on which to base effluent charges unless a single authority is responsible. In federal systems or along rivers spanning more than one country, different states may fix different levels of purity or temperature even though the river passes through each of them. The ultimate confusion can occur when two states sharing opposite banks of the same river fix different standards. In the United States, for instance, West Virginia has fixed 86°F as its top temperature along the Ohio River. But at midstream, the state of Ohio, opposite, allows it to go up to 93°. But with a single authority responsible for the management of the whole system, charges can be both uniform and sensible. They can be based upon a total view of the system and reflect the common denominator of all its needs and opportunities.

The second point follows from this. One of the great problems in dealing with water management, or indeed with any kind of management of natural resources, is that divided authority usually leads to a multiplicity of single-thrust solutions. Engineers will always build dams. Municipalities always want their own sewage plants. Industry always wants tax incentives. But the sum of such policies can all too easily be *less* than the whole.

But a single responsible authority, guided by the most careful environmental studies, inspired by an integrative approach, and given the powers it needs for responsible action, can, like the Ruhr *Genossenschaften,* use all the available policies in the best available combination. There need be no more claim to exclusive effectiveness or risk of technological tunnel vision. Policies can include drainage and dams, cooling lakes and treatment plants, holding reservoirs and blowing oxygen into the water to increase its carrying capacity. An optimum mix of all these techniques can be evolved and their use varied to suit seasons of high flow and low flow, of drought and flood.

In short, the waters can be seen again as creatures of nature, serviceable and friendly to man if he knows the way to secure this cooperation but sullen and dirty servants if this is the treatment that is meted out. And, as any intelligent manager has always known, willing and healthy partners are not only more attractive companions. They are also a far more economic proposition than worn-out slaves.

8 MAN'S USE AND ABUSE
OF THE LAND

The Problem of Wastes

WHEN WE TURN TO man's uses of the land, we confront two main problems of environment and amenity which are hardly solvable in terms of traditional economic calculus. The first is common to all economies and we are by now familiar with it. The exclusion from production costs of the external diseconomies of solid wastes has loaded the land with mountains of garbage which impose unrecognized and unestimated costs on the community at large. This problem is closely linked with the pollutants of air and water and it is therefore probably more sensible to consider it first.

But logically this order of attention does not seem too reasonable. The wastes are the by-products of the way men want to live and the things they want to do. It should perhaps be considered only after we have taken a look at the problems raised by man's new and increasing tendency to have and do his industrial work in built-up urban areas and to extend the techniques of industrial technology to agricultural pursuits.

However, the way in which men live on their territorial units and the multiple uses they want to make of essentially limited space raise problems of so much larger importance than the disposal of wastes that the smaller issue had better be disposed of first.

But *small* is perhaps a misleading adjective. In fact, modern man's greatest contribution to pollution is increasingly taking place on the land. In urban, industrial, and even holiday areas, the problem is the mounting load of solid wastes that must be disposed of. In agriculture, the problems of pollution also include wastes, in particular the 50 per cent of agricultural waste that comes in the form of manure. Thus all of modern man's metamorphoses—as industrial worker, as city dweller, as recreation-seeker, as farmer—leave behind, like the slime of a moving snail, a thickening trail of solid waste.

The figures for waste are already so large that they are hard to grasp. The most detailed statistics are available for the United States. In 1920, the average household generated 2.7 pounds of solid waste a day. By 1970, the figure was 5.3 pounds. By 1980, it may be 8 pounds. If one adds industrial wastes—from mines and factories—the per capita figure is nearer 50 pounds a day.

The various inputs making up this remarkable tonnage must be examined separately since their nature determines how final disposal can best be achieved. On the whole, as we have seen, cities in developed countries manage to get rid of human wastes through their sewage systems without reverting to the horrors of nineteenth-century sanitation. It is also true that a lot of other organic wastes from households —the outer leaves of lettuce, apple skins, pot scourings—are now under better control in the cities than in the past, when they used to be chucked on the town dump or dropped from second-floor windows on to the streets below, there to be eaten by disease-carrying rats or scavenging dogs. The problem today is thus not so much the risk of urban epidemics as the sheer scale of the refuse with which society has to deal.

We have few statistics for the amount of waste rock and mill trailings left behind by quarrying and mining. In the mid-sixties, the United States was reported to be digging out 5.6 billion tons of rock a year, and over half of it was left behind after the extraction of useful minerals in the 8 million acres which in the U.S. are lying in underground mines. The vast tips of South Wales, the mountains of mine tailings that surround Johannesburg are reminders of large-scale mining's cumulative effects. The American figures have obviously been very greatly increased by the growth of strip mining in the last decade—3 million acres of land have been stripped in the U.S. and only one-third of them have received any reclamation at all. All these leavings create special problems. Acid seeps out into underground waters. The land above abandoned mines subsides, leaving houses balancing on the edge of hundred-foot craters, the back verandah high in the air. Worst of all, mine dumps frequently catch fire and go on burning with a slow and enormously expensive obstinacy.

Once again, we have some American figures for the mid-1960s, when no less than five hundred waste-bank fires and two hundred mine fires were recorded.

In the main, municipal systems have to dispose of both urban and industrial waste. The range is enormous, from "wet" sewage, which is potentially a material for composts and manures, on through old, discarded steam engines, broken garden furniture, unusable scraps of metal to the extravagant throwaway items of the modern consumer, which, in the most affluent of all societies, the United States, includes each year 48 billion metal cans, 26 billion bottles, 65 billion metal bottle caps, and 7 million junked automobiles. This process of discarding also appears to be speeding up. In 1960 only about 3,000 cars were abandoned on the streets of New York. By 1970, it was 70,000. The packaging explosion is particularly rapid. In 1958, each American used about 400 pounds of materials for wrapping, bottling, and canning. This may double by 1978.

If this vast and rising tide of solid wastes is not to bury society in the way in which the seven cities of Ur of the Chaldees have been successively buried under the detritus of seven levels of "civilization," two problems have to be solved. The first, and in many ways the most formidable, is to collect the stuff. It has been dispersed through the choices and whims of a billion customers, each carrying away his consumer goods, heavily wrapped in paper or glass or aluminium. None of it is "consumed." It is simply used—for an hour or two if it is a beer can, for a few years as a car, perhaps for a lifetime as a brass bedstead. But at the end of the time it is still there. How can it be economically collected from a billion collection points in homes and factories—let alone from all the roadsides and picnic sites heavily strewn with the bottles and paper bags of unconcerned travelers?

The miracle is that, in the cities at least, the enormous package of material things that modern society tries to carry on its loaded shoulders is on the whole cleaned up. But collection counts for the overwhelming part of the cost. In Britain, for instance, for every £1100 paid out to collect the garbage, it only requires 35 pence to get rid of it afterward. Even so, the usual method—that of putting wet and dry garbage together simply because sorting is too costly—is biologically wasteful since the wet garbage includes valuable elements for compost and fertilizer. Moreover, economic costs are not the only ones. There is no more lethal sound for the would-be urban sleeper at dawn than to hear the bumping and dumping followed by the high-pitched squeal—somewhere between a dentist's drill and an air-raid warning—emitted by trucks collecting and grinding the local garbage.

Disposal, though cheaper, presents its own problems. Unless the garbage is carefully concentrated and properly buried under the right amount of compacted soil, the dumping place becomes a haunt of rodents and insects and develops its own range of unappetizing smells. Sometimes, too, they burn. One near Washington smouldered for twenty years. Landfill areas are therefore rarely welcome to the neighboring population, and finding sites that are neither too expensive nor too unpopular increasingly taxes the ingenuity of local leaders. Peoples' growing sense of ecological balance is already creating opposition to the old practice of finding the nearest swamp or wetland and filling it in. The value of such natural settings for wildlife and for watersheds and microclimates are now felt to be much more important. Bewildered councilors, rerunning for office, find themselves defeated on the grounds of paying insufficient attention to the local needs of wild geese. And if sites are scarce now, what might not be the outlook after another decade of high consumption? To adapt a query first put forward in the sixties, will modern man reach Mars while standing up to his knees in garbage on planet Earth?

Yet there are already a number of ways through or round our mountainous mazes of solid wastes and there are many more that should be submitted to proper research and experiment. If we take first the prob lems of collection, we can begin at the most direct and personal level. The whole weight of social education should be slanted *against* the throwaway mentality. Exemplary fines for anyone actually caught creating litter would speed up the process of indoctrination. Yet there are marked differences in standards of good order between a variety of cultures in already developed, and hence garbage-disposed, societies. This shows to what degree the citizens' sense of responsibility for his surroundings is a matter not only of temperament but of training.

Next, there can be a decisive reinforcement of the present move toward the returnable container. If municipalities or groups of traders in responsible corporations would establish collection points and concede a large enough return on each can or bottle brought back, there are quite enough Boy Scouts or young Komsomols or dedicated adults in most societies to secure a large measure of cleaning up, in part for pocket money, in part for private charity or the public good. Governments could also impose a high enough surcharge on containers to make it worthwhile for most housewives to bring them back. Industries can also play their part. The aluminum companies in America have recently set up their own centers for the collection of aluminum cans. They pay half a cent a can and in 1970 appear to have collected 115 million of them. This figure may still be less than 10 per cent of the cans actually sold but with more centers and more advertising, a much higher percentage can probably be retrieved. And there is some talk of extending the principle to bottles and the scrap recoverable from tin cans. The process could undoubtedly be speeded up if industry-wide regulations were introduced, prescribing types of cans or bottles which can most easily be recycled.

But a great deal of reusable materials—wet sewage, for instance, or massive amounts of paper—are not easily recycled because they are difficult to separate. Citizens cannot be bothered. Separate and more frequent collections make the whole process even more expensive. In most existing cities the retrieval of some value from all the discarded objects of the high-consumption society will continue to come—or not come—not so much from changes in collection but from better arrangements for disposal.

However, in new cities or new extensions of old cities, it might be possible to repeat for garbage what has already been achieved for sanitation. About a third of the cost of cities is already underground—among the sewers and power lines and gas pipes that clean and heat (or cool) the urban dwellers. It should not be beyond the engineers' wit to add a few more channels. Even now sewage pipes are often far from overloaded. Wet garbage, ground to a pulp in a grinder attached to the kitchen sink, could be sent along with the rest of the sewage to the treatment plants. In new cities, sewage pipes could be larger and grinders obligatory. One could also consider chutes down from kitchens to tunnels along which moving bands take off solid waste for separation and reuse. It is curious to reflect how little rethinking of basic urban design there has been since the sanitation revolution of the nineteenth century. Sewers, cables, and storm drains must have seemed quite radical then. Now all too often they appear to represent the limit of the sanitation engineers' imagination. With garbage rising as relentlessly now as did sewage a hundred years ago, perhaps the time has come to think again. France, at least, has taken up the challenge. At Vandreuil, near Rouen, a new city is being designed to be the first "pollution-free" industrial city, with traffic routed underground and pollutants and wastes removed through tunnels.

There are other possible approaches to disposal. For decades now municipalities have in fact been filling in old gravel pits, degraded land, and swampy bottoms and turning them into marinas or housing estates. It has also proved possible to regrass barren areas and redesign them for parks and tree-planting. There is a rising demand for more adequate controls over strip mining, to ensure that, as far as possible, top soil is returned, trees replanted, and damage made good. Indeed, a natural marriage is waiting to be made between strip-mining sites and municipalities overloaded with decomposable wastes. Old underground mines could also be used, especially since most of them are near railways. There are also examples of imaginative new uses for unwanted trash. Near Chicago a whole recreation area with a lake and ski slopes is being built over a marshy pit once known as the Badlands. The clay that is dug out is used, in the proportion of one to three, to make an impermeable cover for the compacted garbage; in the end, the citizens will have lost an eyesore and acquire a landscaped hill, not inappropriately renamed Mount Trashmore.

Incineration is the second most widely used method of disposal. Although municipal incinerators conjure up visions of smelly black smoke and little charred chunks of the daily newspaper floating out of the blue onto the citizens' new white shirts, they can be constructed to high standards of cleanliness. A new model incinerator in Düsseldorf, for instance, services 700,000 people and yields revenues of $3.40 per ton of processed refuse. Industrial users are charged $3.00 per ton to dump their wastes. Steam generated from the combustion process is sold to other city departments for use in space heating; scrap iron recovered from the furnace is disposed of at commercial rates while the ash is sold as landfill or aggregate for such commercial uses as cinder blocks. Osaka has gone a step further and constructed an incinerator which, while meeting strict air pollution standards, not only burns trash to produce electricity but also contrives to burn the city's sewage sludge. In fact, a very wide range of substances can be recovered from incinerator ash. Glass, for instance, can be used in a variety of ways. In Kansas in the United States the glass industry has even constructed several miles of interstate highway using "Glassphalt"—glass instead of gravel as an aggregate—which is reported to be standing up to the test.

There are also a variety of new ways of treating organic wastes in such a way as to rescue from them their minerals and their value as fertilizers. In the Netherlands for the last forty years, at least 30 per cent of the cities' wastes have been returned to the land in the shape of compost and this in spite of a very high use of artificial fertilizer. Finally, there are methods under experiment in which organic wastes are broken down into their constituent elements by distillation at very high temperatures. This process, known as pyrolysis, produces absolutely no pollution since all the distillation takes place in an enclosed vessel. Another advantage is that man-made compounds like artificial rubber and plastic which normally gum up an incinerator give little trouble and a commercially salable gas can, possibly, be produced.

Clearly, many of the techniques used to clean up atmosphere and water and to dispose of solid waste will recapture materials for use and enable a sustained rate of growth to be maintained with much less strain on the earth's resources. One of the ultimate visions behind the development of nuclear fusion is that a "fusion torch" might be invented which would break down all wastes into their constituent elements ready for reuse in what would have become essentially a closed, self-perpetuating ecological system. And short of such grandiose instances of ultimate technology, the installation of modest "car shredders" in most large centers would permit the reuse of virtually every piece of metal and plastic in the junked machine.

One further point in relation to pollutions of the land remains to be made. We have seen how the carelessly lethal use of pesticides, sprayed from the air, has disrupted natural cycles and ecosystems. We have looked, too, at the evidence that artificial fertilizers can contribute to the pollution of streams. At this point, a number of scientists ask the further question: May not the whole development of modern industrialized farming with heavy machinery and a very heavy use of fertilizers represent a dangerous simplification, a tend toward monoculutre, which, being of its very nature far more fragile and vulnerable than balanced, complex ecosystems, exposes mankind to the risk of securing high food returns in the shorter run in return for catastrophic risks of famine later on? May not the thousands upon thousands of square miles under grain on the North American plains prepare the way—to which 1970's corn blight was a first modest pointer—for the possibility that pests will sweep through harvest fields, leaving behind a ruin of rotting heads and mildewed stalks? And, the critics ask, short of such extensive disaster, may not the soil be undergoing steady and irreversible deterioration under the impact of the heavy machinery used to harvest and transport the grain and of the downpour of artificial fertilizer needed to keep up the level of output per acre which high consumption and heavy capitalization seem to require?

They also claim that these evils may be compounded by the urbanization of society and by increasingly industrialized forms of animal husbandry, both of which removed man and beast from their traditional places in the farming cycle. If food is being continuously poured into them while their excrement is simply washed off into the oceans, burned off into the air, or even compacted and buried under Mount Trashmore, a fundamental cycle of nature—the return of natural nitrogen to the soil from which the harvests have removed it—is quite simply broken. The tilth, the humus, the texture of the earth are steadily deteriorating. Given a few more centuries or even decades of this kind of irrational mining of the soil and man may wake up one day and find that he has irretrievably damaged the minute covering of soil that is all that lies between him and the bare rock of the lithosphere.

All in all, the critics put up a formidable indictment. When we consider the planet's still-increasing population, we may even wonder whether the traditional figure of the harvester may not be turning before our eyes into that other image of the man with the scythe—the archetypal image of Death itself.

There are, however, a number of more reassuring points to be made. In lands of reliable rainfall or water supply, some crops have existed in types of monoculture for millennia without significant deterioration of the land. The rice lands of China, Japan, and Southeast Asia have survived although in the past at relatively low sustained levels of productivity. Sugar cane is a particularly stable and productive crop, growing as it does at a point of maximum fertility where air, water, and soil seem naturally to combine in the most productive proportions. In the great grain areas, rotations, varieties of seed, different types of crops planted in alternative belts can all lessen the undoubted risks of monoculture. A judicious introduction of woods and lines of trees as windbreaks, which encourage evaporation and vary the region's land and insect life, has been found in parts of Europe to increase fertility sufficiently to offset any

loss of arable land. In Schleswig-Holstein such practices are claimed to
have increased yields by 20 per cent. Besides, science itself can be used
to recreate complexity even in apparent monocultures. In the great plains
of western Canada, annual variations in seeds and types of plants, coupled with a careful attention to the timing of farm operations, lessens the
risk of mounting attacks by predators or pests on a single, repeated crop.
Moreover, in the whole of North America, an exceedingly vigorous
extension service, backed by a good deal of ongoing agricultural research, provides a background of monitoring, care, and feedback which,
in a sense, puts back into the system some of the complexity that has been
taken out.

On the issue of fertilizer use, some scientists deny that any dangerous
depletion of the soil takes place when natural nutrients are removed and
replaced by artificial fertilizer. Others see a risk but argue about the
degree of damage. As with the problem we discussed on an earlier page
—that of nitrogen running off into the rivers and encouraging eutrophication—the experience of different countries does not suggest a universally deleterious effect. The fact that more grain is being produced on
fewer acres as a result of high fertilizer use *need* not mean that the
cultivated land is being exhausted. The nutrients, once mined, may be
replaced. The whole process may be a genuine jump in productivity like
neolithic man's move from gathered to cultivated seed. In some countries
it can also mean that some marginal land is being withdrawn from
cultivation and can be reforested or held fallow for future use.

Yet there is a case for arguing that there is waste, if not actual mining
of the soil, in the degree to which human and animal excretion are not
put back into the farming cycle. This is the point at which the issue of
capturing natural nutrients and fertilizers from organic wastes becomes
central. Of the two billion tons of waste produced by American agriculture every year, at least one billion is straightforward manures. Unused,
they can lead to pollutive run-off into water courses and to a steady rise
in the costs of waste disposal. At present, the economics of agriculture,
particularly the high labor costs, do not give the farmer any incentive to
return natural manures to the fields. Nor are most municipal systems set
up in such a way as to turn human wastes into useful fertilizers. By
disregarding environmental costs and by interrupting one of nature's
cycles—the passage of nutrients from the soil through animal stomachs
back to the soil—the industrialized nations have contrived to give artifi-

cial fertilizers a price below their true cost.

How to change this possible skew in the farmers' calculations is a complex problem. A tax on feedlot effluents is a possibility. A tax on artificial fertilizers coupled with subsidies for the transportation, through variously adapted sewage systems, of natural manures might ultimately represent a more economic use of the community's resources. What is clear is that the present pattern has grown up empirically as a by-product of unintegrated farm policies and unplanned urbanization. The wastefulness and environmental costs may be tolerable in the short run. But the situation they represent was certainly not planned for the optimum use of land and resources and it is not clear that it represents either the best economic or environmental solution.

Facts of Urban Growth

For millennia, man's main work has been on the land and his settlements have been small, based upon face-to-face communication and in large measure limited by the distances he could go on foot. As late as 1790, 95 per cent of Americans lived in villages of 2500 or fewer inhabitants. Only two towns, Philadelphia and New York, had reached 35,000. In virtually all early settlements—villages or small towns—men occupied family houses and farmed in small communities, the two smallest scales of physical settlement. In one or two countries—one thinks of an ancient capital city such as Rome or of London in the seventeenth century—a new level of urban settlement had come into existence, with a million and more inhabitants. But before industry brought work from fields to urban factories and provided the surplus for a surge of population in the nineteenth century, no one could have foreseen with what speed the whole scale of urban settlement would expand. At present, if we take a community numbering 20,000 inhabitants as the first rung on the rapidly rising ladder of urbanization, well over half the people in developed countries already live in urban communities and over half of them live in large cities with more than half a million inhabitants. In addition, the largest concentrations of population—metropolises of the size of New York, Tokyo, Moscow, Calcutta, and Buenos Aires—where seven million and more citizens have gathered together, are growing at twice the speed of lesser cities. If we were to extrapolate these trends unchanged to the year 2000 we should find over 80 per cent of the world's developed peoples in urban areas. In the largest conurbations London would cover most of southeast England, Boston reach Washington by way of New York, and Tokyo forming a single megalopolis of 30 million swallowing Yokohama and completely surrounding Tokyo Bay.

Moreover, the interconnections between countries and regions, between different communities of learning and interest, between professions and interest groups, all linked by global communication and rapid air travel, are creating, over large areas of the globe, a kind of planetary community or "ecumenopolis" in which, without any spatial contiguity, contact is so immediate, activities so intertwined, interdependence so inescapable that men have started to talk of the "global village."

This kind of unifying, urbanizing world was in large measure a product of the eighteenth- and nineteenth-century creation of a world market. Yet reliance on primary economic pressures and decisions has not proved the most useful guide for building an acceptable postindustrial and largely urban environment.

The first reason we have examined already—the concept of industrial costs which, by leaving out the major external diseconomies of pollution and waste, dumped on high concentrations of people in cities an amount of debris and waste which accentuated the problems already created by the disposal of their own human effluents.

The other two reasons are more complex. The first concerns the economic cost of massive house-building in cities in order to give shelter to a sudden, vast increase in a work force that is becoming urban. All types of economy—market, mixed, centrally planned—have encountered this pressure. Whether the costs are borne privately or publicly the only answer, in the early stages of urban development, is concentration in order to economize on space and materials. In the nineteenth century this need produced the back-to-back terrace houses of a London or early New York. But as building methods and materials become somewhat less dependent upon small firms and manual labor, and industrial materials —concrete, steel—began to be introduced and the infrastructure of sewage and power become more costly, the era of the tenement and the apartment building opened. These structures were often eight to ten floors of separate apartments, jammed together on the smallest space with minimum amenities.

One of the concepts made possible by public control over all urban land in the Soviet Union after 1917 was the joint planning of housing neighborhoods and services. Practical realization of the concept was, however, very often delayed by the pressures of extensive war damage and single-minded concentration on the building-up of industrial infrastructure.

The second reason is more far-reaching. It is rooted in the fact that land is inherently limited. Sooner or later, as population, urbanization, production, and recreation all rise, alternative uses for particular pieces of land begin to compete with each other, even in societies as vast as that of the United States, which began its independent life with 5 million people in an open continent—India at the same point in time already had over 100 million people in a smaller area. Once land scarcities appear— and nowhere have they done so more drastically than in the center city —undue reliance on price signals alone can lead to types of land use that grossly contradict amenity and human values.

Let us look briefly at the way in which the unresolved pressures of alternative use have formed natural patterns of human settlement and spatial location. A convenient starting point is the urban concentration which followed rapid industrialization in the nineteenth century. No population then grew by more than 2 per cent a year. Today, growth is nearer 1 per cent. But this relatively unexplosive expansion of people has been offset by another factor—a steady decline in city density. Between 1940 and 1960, for instance, the proportion of people living in the center city fell from 54 per cent to 27 per cent in Stockholm, from 71 per cent to 37 per cent in Toronto, from 77 per cent to 41 per cent in Madrid. These percentages are somewhat distorted by the number of rural migrants who have moved directly to the suburbs. Even so, the peak density in developed cities seems to have been passed by 1870. In all affluent societies, the major shift of population has been not to the core of existing cities but to vast areas of suburban growth, made possible first by the railway and then by the car.

It has also been accompanied, in the wealthiest of these societies, by the appearance of a second home in the country or near the seashore for weekends and vacations away from the stress of the city. Thus the impact of urbanization on a society does not lie simply in the growth of numbers. It lies in the constant increase in the claims made by urban people upon the available land within reach of metropolitan centers.

In the Netherlands, for instance, the most densely populated country in the world—and one which, moreover, has had to "invent" much of its land by dredging it out of the sea—they still only have 14,000 square miles to live on. If open-ended, uncontrolled urban and suburban growth were to be the basic pattern of their development, the Netherlanders might become the first "spread-nation" in the world, the first in which virtually all nonurban activities had been buried under the remorseless advance of houses, industrial plants, and high-density mechanized and factory-type agriculture.

This is, no doubt, the extreme case. But the risks which have induced the Dutch to introduce, by way of a National Physical Planning Act, passed in 1965, a comprehensive approach to land use in their congested country are beginning to move to the center of environmental preoccupations in all affluent lands. The conviction is growing that a hit-and-miss, free-for-all, totally uncontrolled, laissez-faire approach to the use of increasingly scarce resources of land only leads to a series of interlocking environmental disasters. Uncoordinated urban sprawl in big conurbations drains away life and vitality from intermediate cities. The urbanization of high-quality farmland wastes an unrenewable natural resource while regions of great natural beauty are lost precisely when they are becoming essential for the future recreation of more leisured societies. The Dutch believe, for instance, that work will occupy only 42 per cent of the citizens' time by the year 2000. The Russians recently charted all their mountain, river, and coastal areas in order to plan for the greater recreational needs a shorter working week will entail.

If no effort is made to achieve a survey of alternative land uses and on this basis to evolve a broad plan for all the variety of claims modern man makes upon his inherently unexpandable resources of land, the result will be not "civilization," not "urbanity"—words often all too incongruously associated with city living—but a monumental environmental mess.

There are two elements in a rational survey of land use. The first is composed of the basic needs of human beings. The second lays down the opportunities and difficulties inherent in realizing them on the available physical terrain. There is constant interplay of the most intimate kind between the settlements man builds and the spatial constraints and opportunities offered him by his plains and rivers, his hills and coasts. We can look at the living space available in modernized countries under four broad headings—the wholly urban landscape of the city core which, in one way or another, can claim to be the central node of a wider urban area; the surrounding landscape of suburbia—part built, part green but entirely dependent upon its urban connections. In a number of conurbations, this suburban belt may include a number of concentrated urban centers coexisting in more or less built-up areas stretching over hundreds of square miles.

Then comes the man-made landscape of farming and forest. Last of all, is the wilderness—not the wilderness of derelict land and abandoned mine pits but the world of seascapes, lakes, and mountains, as far removed as possible from the busy activities of man.

It is clear from this categorization that man's options in developed lands are very considerably restricted by what he has already done and built. But with greater pressures ahead, the options will be even smaller if he does not adopt saner procedures for land use and do so in time.

How do man's built-up and natural environments and their interactions meet or frustrate his human needs? These may seem straightforward questions. But the difficulty lies in arriving at satisfactory definitions of what human needs really are once we leave behind a basic biological minimum. We know that people need to eat, to be housed, to be healthy, and to grow up in some sort of family or clan. But beyond this basic minimum, we enter areas of great cultural difference and also of a considerable ignorance of the facts. We can guess that few human beings are born to be happy in complete idleness. But we do not know much about optimum and minimum amounts of work. Tribal families in subsistence economies work hard for not more than three to four months a year. There is evidence—from the number of saints' days and festivals—that medieval man may not have clocked up much more than 190 days a year. With the onslaught of the Industrial Revolution, when for a century men worked for seventy to eighty hours a week in inhuman noise, dirt, and stench, some deeply ingrained habits of self-entertainment must have been profoundly disturbed. And the cities built to accommodate this pitiably exploited work force were not designed for anything but sleeping and sweating it out.

Yet now that the formal work week is steadily falling, recent surveys in a number of industrial cities do not give any very clear picture of the degree of leisure people seek. In the sixties, in Britain, for instance, one man in ten had a second job and one in three did overtime. Income was clearly more inviting than more hours of leisure.

How much of the time used for leisure is spent outside the home is also difficult to judge. Some surveys suggest that not more than one citizen in ten takes part in any kind of organized sport. Television keeps people indoors. But the motorcar takes them out and clearly it becomes the prime instrument of leisure when young families are formed and weekend visits to picnic sites and seashores produce the endless traffic blocks on all major roads leading back to big urban settlements on Sunday evenings.

That urban man obviously likes to get away to nonurban surroundings is confirmed by the phenomenal growth in international tourism. Since the last war, there has been a remarkably large increase in the number of long-distance journeys during the period of what is now a fairly general two-to-three-week holiday. The United Nations reports that between the mid-fifties and the mid-sixties, the number of tourists arriving in some sixty to seventy countries rose from 51 million to over 157 million. These figures also tell us something more about modern man's less basic needs. He wants sunshine. He seeks the sea and the mountains. He is drawn by old and beautiful cities. He positively swamps areas where—as in the Mediterranean or the Alps—he can get a combination of these things. Tourists in Greece have trebled in the last ten years. So have visitors to America's national parks. People do not cross the oceans and the continents to look at industrial plants or inspect uniform suburbia. The hunger lives on for beauty and for natural things.

Can we say how strong this hunger is? Will an annual escape to the isles of Greece or—more likely, the towering hotels and crowded beaches of the Costa Brava—be enough to keep a citizen's aesthetic budget in balance? That he can survive inhuman and degraded conditions is certain. How, otherwise, would so many wise, gifted, and dedicated citizens have found their way out of the vile slums of nineteenth-century Europe or America? But for every child who succeeded, there were dozens who succumbed to thieving and killing and starving. One of the most terrible pictures of nineteenth-century London is a reporter's tale of a youth of twenty-one trying to drown himself in the Thames and fighting his way back through the black mud into the water against all the efforts of rescuers to pull him out.

Besides, we do not fully understand the longer-term results of extreme cultural, ethical, and emotional starvation. When we remember under what continuous stimulus of natural variety—of color, of scent, of sound and light and touch—the first men began to develop their imaginative grasp upon living reality and feel their way toward fully conscious and creative humanity, we may wonder what will be the result of a continuous adaptation of human existence, over centuries, to towering buildings, concrete walls, personal isolation, darkened skies, roaring traffic, raucous noise, polluted water, and dirty streets. Such an urban environment might begin to produce human beings whose very ability to survive in such conditions could mark the beginning of a retreat from realizing their full potential. The remarkable, the resilient thing about man is his ability to adapt and survive. But some adaptations become deformations.

These are new questions because it is new for man to live in unrelieved urban or suburban man-made surroundings and because most city dwellers in developed countries have, so far, found avenues of escape —to the sea, to the lakes, to the mountains. The issue will become more critical as the developing world undergoes its urban revolution and by 1980 adds to its city populations the equivalent of the entire present population—roughly a billion people—of the developed world. Pockets of urban degradation in affluent countries provide us with some foretaste of what could become man's greatest environmental risk—spreading urban misery, city quarters of unrelieved ugliness and squalor in which the imaginative life of young children may be as systematically starved as their bodies are undernourished, in which they are cut off, by early deprivations, from their full cultural and even human heritage. If, in spite of such destructive environments, the flame of life still burns brightly in so many run-down, squalid slums, it is again and again because in spite of all the poverty and restriction the child's sense of belonging to a living supporting family, neighborhood, community has been nourished by the religious or ethical tradition of the local culture.

Here we are on much firmer ground in defining basic human needs. Most human beings are not solitary. The degree of desired closeness and physical contact varies, of course, from culture to culture. Each of us carries round a kind of minimum envelope or bubble of free space which we want our neighbors to cross only by invitation. The difference between a cool handshake and an effusive *abrazo* may be quite enough to complicate the coexistence of two different ethnic groups in the same city quarter. Yet a community in which different families and individuals can meet and get to know each other face to face, band together for common enterprises, support each other against outside interventions, and experience a sense of the more profound significance of their daily living has been shown to be a need among living creatures ever since man emerged from the primal groups of herd and pack.

This sense of community is more than familial or personal. It is based upon a closely related set of institutions and needs—schools, churches, corner stores, clubs, and pubs—which embody participation in the wider society. The largest of all modern city builders—the Russians, who constructed the phenomenal total of nine hundred new cities between the wars—embody this principle in their physical plans. In creating their smallest unit, the microdistrict of 8,000 to 12,000 people, Soviet planners try to project the needed number of schools—primary and secondary— the clinics, food stores, repair and dry-cleaning centers, public places, and small gardens appropriate to a community of this size. A cluster of such units making up perhaps a block of 25,000 to 50,000 people would have, in addition, public offices, bigger parks and shopping centers, theaters, restaurants, and other buildings not in continuous use. As size goes up, so, in theory, does the elaboration of services, department stores, hospitals, centers of education and entertainment.

This pattern is coming to be far more widely accepted in city planning. Provided the city's transport grids can serve as a network for such "village" units, federated in a wider urban system of larger variety and scale, the result can give what modern city dwellers appear to want— a sense of rootedness coupled with a large range of choice. The small community protects them as children and cherishes them in their old age. But the linking of the "village" neighborhoods to the wider city will give adults in their years of working and parenthood access to a variety of responsibilities, employments, and pleasures and, hopefully, allow them to offer their children a variety of possible futures and careers.

At the same time, the city's openness to its natural surroundings will give its people the chance of regular escape to the nonurban world. These are the ranges of choice that draw people to the city, these are the magnets that provide, along with jobs and markets, the "sucking pull" of the big metropolis. Much of the malaise in contemporary society comes from the degree to which, when people reach the urban regions, stability and choice, security and variety are precisely what they do not find.

The Center City

However we define the basic needs of urban man, he is unlikely to find any city designed in such a way as to satisfy them. Integrated urban planning for all the varieties of human needs and uses has been the exception. The land has been dragged along in the wake of industrial and technological change and rapid population growth. Economic decisions have tended, until recently, to influence a very large part of the shape and texture of the settlements. This fact is due to the role of private decision-making in market economies and to the overwhelming need for economic expansion in the centrally planned systems. The result is an arrangement of space which only partially satisfies man's basic needs.

Beginning with the center city, we will take for granted the cleaning-up of air and water, the better disposal of wastes, and the role of the automobile since they have already been discussed in some detail. But the point can be made here that a form of pollution we have not discussed —that of noise—is intimately bound up with the scale of machine use and motorized traffic, private and commercial, operating in a city. All the arguments for underground garbage disposal, for emissionless electrical motors, for pedestrian precincts, for belts and oases of parkland within cities are enormously reinforced by man's limited tolerance for noise above a certain decibel rate.

Of all forms of pollution, noise is perhaps the most inescapable for the urban dweller. It pursues him into the privacy of his home, tails him on the street, and quite often is an accompaniment of his labor. We do not begin to know the price we pay in impaired hearing, in enervation, in aggravated hostilities, and nervous tension. But scientists report that when animals are made to listen to noise "they grow sullen, unresponsive, erratic, or violent." May not the same be true of us?

It is true that tolerance of noise varies with culture and conditioning. A Swede might not be comfortable in Naples. A farmer might find a steel plant utterly intolerable. Some people seek out noise, as is evidenced by discotheques and rock-music concerts, where the decibel levels have been found to be clearly dangerous. Others find the sounds of the city and the marketplace sheer music. Still others encase themselves in sound to protect against sound, thus creating more noise.

But almost all would agree that the noise of jets in takeoff, the air hammer digging up streets, the unmuffled sports car or motorcycle, the roar of twenty-ton diesel trucks are an intolerable burden on our ears and our whole nervous system. The lack of discriminate planning, which has put airports near central cities and has intermingled industrial and residential zones, has caused much of the trouble. Here we seem to be learning. But the prospect of sonic booms remains and with it the danger that—apart from their impact on man—some small animals and insects which rely on their delicate hearing for survival will be badly affected.

Only recently has attention been focused seriously on the technology of noise abatement and the design of strict and enforceable regulations. Much can be done to baffle unavoidable noise. Building codes can require soundproofing in public places, and builders who construct paper-thin inner walls and ceilings can be penalized. In some countries, successful battles have, happily, been fought and won to check the continuous playing of canned music to captive audiences, and in some parks and playgrounds private noise is regulated. But in the last analysis only self-discipline and respect for one's neighbors can lessen the blare of late-night radio or the wailing of transistors on public lakes and beaches.

Indeed, this whole area of the electronic reproduction of sound profoundly illustrates the double-edged quality of so much of man's technological inventiveness. Few people will question the almost unalloyed advantage of extending to the vast mass of the people the ability to hear every sort of music, every kind of play, poetry, and information. It is in many ways the least pollutive and wasteful form of man's entertainment. Privileges once enjoyed only by aristocrats—to hear or watch the world's finest artists perform—are now the common heritage of millions. Perhaps especially for the old, the range of interest and stimulus, unobtainable in any earlier society without long and difficult journeys, is an incomparable benefit.

Yet there is a price—the price of growing, privately produced, indiscriminate noise. This is the first epoch in which an immediate neighbor can have a complete brass band playing in the kitchen. The new opportunities demand new sensibilities and restraints. All too often, they lag far behind the growth of sheer din.

Noise is one element in the wider issue of beauty and amenity in central cities. Here we must distinguish between ancient cities with beauty to preserve and modern cities which have yet to create it. In beautiful cities the chief problem is to care sufficiently and efficiently for the urban heritage. On balance this seems to be better done when public authorities have the necessary control over land use. One of the most remarkable achievements of this kind, against the worst of all possible odds—the nearly total destruction of war—was the postwar reconstruction of Warsaw with a respect for the past, with a loving attention to detail that restored to the citizens not only their familiar townscape but their sense of historical continuity and heroic resistance. Another remarkable achievement of restoration in a great urban work of art is the renewal and, indeed, the enhancement of Peter the Great's magnificent center of Leningrad. Where, however, city authorities and planners have less control over the disposal of urban land, some tragic collisions of interest can destroy a city's heritage.

There can in fact be a fateful connection here between the private land market and the destruction of urban values. We have already noticed the nineteenth-century treadmill effect of low rents requiring overcrowding to create a profit and the resulting profit forcing up the value of land in urban areas. These values, created very largely by the pressures and needs of the community, can, when netted by private developers, create a new kind of treadmill—the need to secure very large returns on the limited amounts of land available in central areas and hence the development of the huge skyscraper with all its destructive influences on the urban landscape.

It is destructive first because it can put every other traditional building out of scale. As late as 1945, London's skyline was one of its glories. The tall spires of Wren's churches, the balanced and monumental dome of St. Paul's, above the wide flowing arc of the Thames, the great parks, the green rise of Hampstead, and the hazy outlines of the North Downs gave London well into the middle of this century the proportion and splendor Canaletto had painted two hundred years before. This beauty has faded in only twenty-five years. Commercial and public buildings of monumental size and ugliness have risen haphazardly in all parts of the city, like a scattering of tall pepper pots on a carelessly laid table.

We need not say that all buildings of thirty floors and upward are visually outrageous. Carefully grouped, in balanced relation to different planes and levels, they can have a stimulating effect, particularly in new cities. But, in new cities or old, skyscrapers built by the chances of zoning controls or land purchase are virtually certain to be haphazard intrusions on the city's human scale. Built all altogether in long lines along canyon-like streets as in Manhattan, they make an environment for ants, not men.

Their defects are more than visual. High-rise towers have proved a disastrous experiment in urban dwelling. They give many of their occupants acute uneasiness. Some people arrange their furniture so as to avoid any view of the vertiginous plunge from their thirtieth-floor window. For mothers with small children, they present insoluble problems of play and supervision. The elevators become places of dirt and danger. The wholesale bulldozing of little streets and houses to make way for them destroys delicate networks of service and friendship which are simply not re-created between different floors in new apartment houses. The ground areas between the towers, which were supposed to provide needed air and space and greenness, can become windy deserts below vast buildings which tunnel the weather down their vertical sides as do mountain ranges.

Some town planners even maintain that the claim made for high-rise dwellings—that otherwise even more little houses would be scattered over the countryside—is not borne out by economic or spatial necessity. In a number of cities, areas of similar size, with alternations of four- to eight-floor blocks round enclosed gardens and courtyards, can house virtually the same number of people and provide the intimacy and security which parents in particular look for in an urban home.

It is surely significant that two of the most densely populated countries in the developed world, Britain and Holland, are reconsidering high-rise living. In the chief towns going up on land reclaimed from the Zuyder Zee, the bulk of the housing is in single houses with gardens. Recently, the Greater London Council removed all future high-rise dwellings from its drawing boards—a not irrational response to the discovery that 80 per cent of their tenants were miserable in them.

But perhaps the worst urban evil represented by high-rise commercial buildings in a number of developed cities is the evidence they give of resources diverted from the most fundamental of all environmental urban evils—festering slums and hopeless ghettos. This is particularly flagrant in any city where no taxes are paid on large office blocks and they stand empty, earning capital appreciation while the slums survive or where tax concessions and even tax havens are available in relation to commercial building while public and private construction of homes for the poor lag behind.

Many of the new constructions—high-rises, apartment blocks, even handsome new streets and layouts—represent a further degradation of living conditions for the poorest citizens. In the name of slum clearance their streets are knocked down. And what is rebuilt is far beyond their modest incomes. They move on, doubling up in older houses and spreading urban blight still further. As early as the mid-nineteenth century, a rueful London verse ran thus:

> Who builds? Who builds? Alas, ye poor
> If London day by day "improves,"
> Where shall ye find a friendly door
> When every day a home removes?

A decade later, the ruthless Baron Haussmann, carving out his celebrated boulevards in Paris, scattered the dispossessed poor to garrets in nearby slums, greatly increased the density of population and with it the ravages of tuberculosis. Yet, in the next century, the first large-scale experiments in slum clearance or urban renewal in the United States had some of the same effects, increasing the land available for developers to produce apartments at higher rents and squeezing the urban poor into other run-down neighborhoods which then deteriorated still further.

The high value of urban land is not only responsible for the antisocial aspects of high-rise buildings. It also accounts for the fact that too many modern cities are wildernesses of stone. Again and again, we find that the great parks, vistas, and open spaces that make London or Rome or Paris such targets for tourism are the legacy of earlier royal or aristocratic initiatives. The treeless city among everlasting concrete trenches and barracklike agglomerations of brick and stone reflect in market economies the overvaluation of every inch of urban land.

In some of the centrally planned economies, comparable conditions reflect the extreme pressures of forced draft industrialization and, more recently, an expensive rebuilding after war. In wealthy lands, these city deserts are made more unviable by the alienation bred of spectacular poverty of some city quarters and relative affluence everywhere else. The effect is particularly brutal if the poor citizen is cut off not only by poverty but by a different ethnic or cultural background.

Usually, in developed countries, the cause of these pockets of delapidated housing, poor services, ugly surroundings, dirt, and disease is basically the same. In a relatively short time, a wave of rural migrants has poured into the city with few skills, little money, and no urban experience. They have taken over the run-down building of earlier dwellers who have now improved their incomes and, as often as not, moved to a suburb. In ninteenth-century America, for instance, the migrants came from Europe. As they improved their standards, the form of American assistance to housing—mortgage guarantees—both before and after the Second World War enormously increased the movement outward to suburban single-family homes. Since 1950, four million houses have been built for people whose average income in 1968 was over $10,000 a year. But into the vacated areas there came—in all the northern cities —a massive black migration from the rural south.

A similar movement away from rural poverty brought Jamaicans and Asians to England and Algerians to Paris. Some of the same problems occur even when the rural migrants are of the same culture and stock. Sicilians have moved massively to Milan, country dwellers have streamed into the great Soviet cities—Leningrad and Moscow grew from 2 million to 7 million between 1932 and 1962, in spite of administrative attempts to stabilize the size of the cities. The consequences have tended to be the same—a pileup of the least skilled in the worst quarters and a desperate need for new and heavy expenditure in urban infrastructure, schooling, and housing. These were major Soviet preoccupations once the immediate damage of war had been made good. In Britain too, although urban housing compares well with that of any other country, in some recent surveys it has been suggested that at least a million families are ill-housed and below the poverty line. Poor housing and lack of home ownership have continued to be a source of unrest and dissatisfaction in Italy's industrial towns. In the United States, so far, the funds needed for a full-scale and radical elimination of ghetto dwellings have not been made available. In fact, only about two-thirds of the estimated annual need for subsidized housing for the poor is being met, and meanwhile the housing stock deteriorates further.

There can be no doubt about the center city's chief environmental priority and it is not one that the unaided market will supply. It is the rebuilding and rehabilitation of all remaining slums—by public investment, private inducements, rent subsidies, tax rebates, and all other appropriate policies. But this priority only indirectly deals with the problem of restoring the city's role and dignity in the wider metropolitan context. Unless cities still command such inherited beauties as London's parks or Paris's Place de la Concorde or the whole Vatican—St. Peter's —Castel San Angelo vista in Rome, they tend to lose any power as cultural magnets or expressions of urban magnificence and civic responsibility. Manhattan's appeal as a center of the arts is perhaps the exception proving the rule, and even here, the beautifying role of Central Park and the river fronts should not be forgotten.

There needs to be a dual approach to the problem. The first can be accomplished in the city itself. The second depends upon its relations with surrounding suburbs and upon the planners' ability to reduce the pressures which arise when a center city has to serve too large an urban region. Within the city, the aim must be to try to build up among and around the modern high-rise citadels new neighborhoods built to a human scale, where people of different occupations and classes can live together in distinguishable communities—like New York's Greenwich Village or Rome's Trastevere—walk to work, and keep the city alive when the offices close. The Barbican area in London is being planned to fulfill this function.

The offices themselves can, as it were, rejoin city life, particularly after office hours, if, as in parts of Manhattan, they are built only on condition that their street floors contain theaters, restaurants, arcades, and generally agreeable access for the city dwellers. Parks—as in Amsterdam—can be introduced to give the variety and freshness of greenery throughout the central area, small square parks, long thin parks, winding through the big blocks, "hanging" gardens between tall buildings, fountains in city squares, flowers tumbling out of window boxes and hanging from lamp posts. Traffic-free areas—or days—can be introduced so that, say, during summer lunch hours people can come out and picnic on the streets and wrest back to social purpose the dehumanized concrete and asphalt realm of the motorcar.

It was said of the capital city of the Sungs that no street was without "the sound of water and the scent of flowers." Those who have visited Hankow recently speak with wonder of the sense of greenness and spaciousness in what is now an industrial city of nearly a million people. The transformations are thus possible but they demand in market economies a far greater public commitment to the ideal of urban excellence.

But the restoration of life and dignity to the center city also depends upon its links with the suburban regions which surround every major city and are the fastest-growing areas of human settlement. Yet for all their obvious drawing power, they cannot be said, any more than the city centers, really to be satisfying the needs and hopes of urban man.

Suburbia

Nobody intended suburbia. It began and has in part continued as an escape from the dirt and pressure of the modern industrial city or simply from the city as such. It is therefore not wholly surprising that what has been basically a random but self-reinforcing process has not, whether for work or leisure or community or even contact with nature, provided wholly satisfactory answers.

Take work first of all. As the suburbs spin outward and one city's spread is only stopped by the next city's sprawl, work tends to involve a longer and longer journey. It is true that light industry and services follow the commuters outward—with the incidental effect of filling up all the green spaces they had hoped to enjoy. But this fill-in has not simplified the commuting pattern. Many workers in the suburbs continue to rely on the center city for employment. Yet a number of workers living in the city are quite as likely to be traveling outward for the new opportunities of suburban work. Either way, enormous amounts of commuting time have to be absorbed. If workers drive themselves to work, the sheer loss of time, the fatigue, and strain are far indeed from the leisure and variety of which the car is supposedly the symbol. Yet public passenger services have tended to run into financial difficulties which reduce maintenance and comfort and increase fares.

In any case, by car or train, there is a certain disamenity, not to say insanity, in spending two to three hours every day in crowded trains or on crowded roadways. If, in addition, the workers are putting in overtime, they can end a work week not noticeably different from the seventy-hour grind of the mid-nineteenth century.

At higher managerial levels, it seems a general rule. In that case, family enjoyments can shrink to a briefly shared weekend, and when children are young, there is a considerable risk of marooning the beleaguered wife. Clearly, one can become acclimatized to the routine—although the costs to lungs, nerves, and family harmony may be cumulative. Once again, may we not confront here an adaptation which is in essence an injury and a loss?

Or take the interrelated issues of leisure, nature, and the need to encounter beautiful things. We should not exaggerate. For millions of suburban dwellers, nature and beauty come in small and much-loved packages of back garden in which prize dahlias are grown and dedicated labor lavished on grass and shrubs and flowering trees. It is not all a wasteful and ridiculous struggle with crab grass. And as wealth goes up and acreage increases and handsome houses, displaying every style from mock Tudor through French baroque to all-glass modern, are reached up winding roads lined with trees and neat white fences, it is hard to believe that the life-style suggested by these well-tended exteriors is entirely divorced from leisure and beauty and natural things.

The difficulty here lies with the concept of community. We can look at it in two senses. The first is the small local neighborhood of shared friendships and interests. The second is the larger community of the total interdependent urban area of which the local suburb or commuter town is a part. Both can be ill-served in the modern conurbation.

Take the local community first of all. Its vitality can be doubly drained. Too many of its inhabitants are basically interested in matters outside the settlement—which may be, in any case, very ill-defined in their mind's eye. Local needs, responsibilities, and possibilities—better layout, new schools, road plans, encroachments, the acquisition of open land—may not rally enough citizen support because the community is too diluted to excite it. But this local thinness can also accentuate the tendency for most suburbanites to look for their major jobs, interests, cultural needs, and professional entertainment to a city center whose services and transport networks become too overcharged to carry adequately such an array of tasks.

Yet the center city is not necessarily better off. Sometimes it is a question of funds. If suburban communities cash out of the costs of maintaining a city center upon which they depend for work and for important cultural stimulus, they, in fact, cash in on the benefits and out on the price. This is particularly serious in cities where some sectors are run-down and present particularly heavy costs of rehabilitation. At the same time, a large suburban belt, particularly if it sets up ethnic or social barriers to the outward movement of poor people, can condemn the inner-city family to a life that is wholly divorced not only from beauty and natural surroundings but in some cities from desperately needed opportunities for work.

In any case, as the suburban belt grows, the inner-city tends to spread, absorbing nearer suburbs into a solid built-up mass. On the outer fringes, open country and useful farmland recede. At last the wilderness itself may be under threat. A sense of "boxed-inness" can even overcome the wealthy suburbanite on a seven-acre lot, so he buys a villa in Antigua or Majorca as well. For the slum child, the trap closes on what is one of the most inadequate ecologies ever designed for living things.

Can anything be done about a process which appears to follow in the wake of every movement toward urbanization and is accentuated sharply by the coming of the motorcar? Are we dealing here with irresistible technological forces which condemn us to a poor and undernourished urban existence—between commuting suburbs and deteriorating city cores? The answers can be given not so much in terms of theory but from the experience of countries which are making the attempt to canalize and direct the urban flood.

Among the centrally planned economies, Romania's rapid industrial-ization and urbanization in the last twenty years gives us some interest-ing ideas. Among mixed market economies the Netherlands should be looked at, not only because of its rating as the most densely populated of all developed countries but because it begins to look like putting into effect the most systematic and all-embracing of national environmental plans.

The first point that both approaches share is a more or less integrated approach to the question of land use. The Romanian territory is rela-tively dense. Its 20 million inhabitants live on just over 237,000 square kilometers and of this 60 per cent are mountains, hills, and plateau country. The whole Romanian territory—with its 20 million inhabitants —has been carefully mapped, its natural regions, lines of communica-tion, soils, forests, and areas of settlement assessed in a national inven-tory. It is within this planned context that a rapid rate of industrializa-tion over the last two decades—of over 13 per cent a year—has brought industry up to the level of providing over half the country's annual product. During the same time population growth in general has not been very rapid—about 1 per cent—but urban population doubled.

It is at this point that we encounter a vital consequence of a careful control of land use. The Romanians set out *not* to create an oversized dominating center city. Between 1948 and 1968, Bucharest grew by only 400,000—from just over a million to just under a million and a half. Its share of the country's urban population actually fell—from 28 per cent to 20 per cent. Meanwhile, economic growth was fostered in other centers. Towns of 100,000 and more grew from 2 to 12, of 10,000 to 50,000 from 60 to 124. Small towns, particularly in regions set aside for water storage, hydroelectricity, tourism, and recreation, hardly increased at all. It is clear from this model that the overcrowding of central urban regions is not an inherent result of the modern technological society.

However, Romania began to practice comprehensive planning when nearly 40 per cent of its peoples were still working in agriculture. The possibilities offered by its type of strategy are of more immediate use to societies in the process of modernization than to those which have already more than a hundred years of industrialization behind them. In these conditions the experience of Holland is more immediately relevant. There, as a consequence of its National Physical Planning Act, the most comprehensive approach to the problem of controlling urban sprawl is to be found. Elected local and regional authorities are responsible for detailed plans which citizens may inspect, accept, or contest. A number of New Towns are being built to take the strain off the biggest urban concentrations—the ten almost continuous urban regions, which include Rotterdam, Amsterdam, and The Hague, the so-called Randstad, where a third of the population live on one-twentieth of the land. These New Towns are based on neighborhood units and include experiments in traffic-free areas, sunken roads for through traffic, and green centers for civic amenities. Nor is this policy confined to the newer towns. Amsterdam had added 1000 acres to its 5000 acres of civic parks in the last two years and plans to bring the acreage up to 7500 by 1975.

Above all the authorities can plan for the Randstad as a whole. This does not take away the rights and powers of smaller bodies. It is a delicate balance between general standards and regional coordination on the one hand and on the other the details, quirks, and delights of small neighborhoods planning their own squares and painting their own houses. In fact, it is this overall control that makes local variety possible and prevents in market economies the single-thrust, overunified, most cheaply mass-produced buildings and layouts all too often produced by the private developer controlled only by the need to beat rising land values and who makes his investment profitable in the narrowest and speediest sense. But it must be said that under pressure any society, market or planned, is likely to produce uniform mass-produced buildings if housing needs become paramount and time is critical.

Britain, like Holland, has experimented with New Towns to take the strain off the big cities. They are linked by good if traditional rapid transit systems with the center city but in theory they are supposed to be self-sufficient enough to take the load off the center's facilities. However, neither in scale nor variety have they developed beyond dependent communities still, in essence, part of the commuter belt. The same kind of problem has, incidentally, emerged round Stockholm where a New Town like Vällingby, built at thirty-minutes' commuting distance from the city and intended as a separate entity, still has over half of its wage earners working in Stockholm.

These shortcomings have led to new thinking. In the last decade city planners have begun to wonder whether their earlier concepts, although innovative, may not have been insufficiently comprehensive and radical. They ask whether they may not have been aiming at better versions of the same thing, at improvements on the old, fixed urban structure of concentric circles, a central core surrounded by subordinate districts. Now should they not think about cities in new terms—of growth rather than size, of mobility, of alternatives and choices rather than a rooted environment?

A radically new pattern begins to emerge. The basic principle is that if the megalopolis acts as a giant magnet, drawing greater numbers of people and square miles into itself, then it must be met with an effective counterforce, diverting the pressure of more population and the crush of their artifacts away from the original center.

Since the urbanization of man is a reality, the countermagnet can only be one or more cities (either brand new or existing ones developed in accordance with careful planning) that lie close to but outside the existing urban magnetic field. And the countercity must be of sufficient size, with sufficient industry and amenities, to act as a counterbalance; otherwise more varied employment, education, entertainment will still pull the commuter into the "great wen." This is the thinking underlying the construction of a new urban center at Milton Keynes in Britain and the development of a regional authority for the whole basin of the Seine in France. Perhaps the most systematic attempt yet made to make a coordinated plan for a metropolitan region as an area of pressure and counterforce appears in a project drawn up in the mid-sixties for the Detroit area in the United States. The initial impetus was given by the local power company's long-range research into where it should plan to lay its power lines. A complete inventory was taken of trends in the region—growth, movement, density of population, sales of farmland, lines and modes of traffic, length of commuting journeys, concentrations of industry—all extrapolated to the year 2000.

This led to an examination of Detroit's regional field of force and its place in wider patterns of movement which crisscross the region, linking it in every direction to other areas of the United States and Canada. Four or five major criteria were chosen—the siting of urban centers, major concentrations of industry, harbors, airports, educational and research centers—and were then tested against such alternatives as density of traffic, speed of movement, traveling time.

Using this information, a complex planning operation was initiated, each stage increasing the detail examined and eliminating the least workable solutions of the preceding stage. The process finally converged on one solution that, in terms of cost, practicality, amenity, and convenience gives the best answer for the entire region's future—the construction of a completely new countercity, Port Huron, at another "natural" intersection of movement and traffic. This would permit both cities to flourish and would not exclude the expansion of Toledo as a third center at a later time.

What can we learn from these various experiments? Man has long been a city builder, a creator, and he has not stopped being so in our era. The difference is that where in the past he learned to achieve this with order, he does it now in all too many countries in a disorderly way. It is therefore important to understand in which directions the new sense of need for urban order is leading the modern state. Some governments are now experimenting with metropolitan control over the whole conurbation to allow pressures to be relieved, costs to be shared and the optimum arrangements to be made over limited space for all urban man's variety of activities. In another range of experiments, particularly in New Cities, public ownership of the land gives the city planners a much greater range of choice simply because they are not battering their way against very high land costs in center areas. The Dutch are extending the practice in existing cities. Two-thirds of Amsterdam, for instance, is owned by the City Council. Those who feel such extensions of public ownership simply add to the cost citizens have to carry need to remember that one way or another urban costs have to be met. The people are there. They cannot be dumped all over the landscape. If there is no planning, a few developers may make great fortunes, the mass of middle-income consumers will get good homes together with commuting and inconvenience, the poorest will be left out. If totally free market systems produced good cities, we should have them now. It is because the market is a limited tool for dealing with *collective* needs and systems that city planning and a measure of public financing are indispensable parts of a decent human environment.

Yet urban planning should be simply part of a much wider approach to the national territory as a whole. It attempts, on the basis of the present distribution of the people, on industry, on history, on climate, rivers, hills, and soils, to make the best, most efficient, but also most beautiful use of a land area which is seen ever more clearly to be limited, to impose choices, to need solutions, to be profoundly destructible unless the twin needs—of overall standards and local variety are kept in some kind of balance. In a sense, a national plan is ecology practiced at the country level, a sorting out of habitats and environments, an understanding of special niches, a creative reaction to forces of dynamic change, a rejection of single-thrust development based upon a purely economic calculus, a search for patterns which satisfy a wider variety of human needs. None of the plans is perfect. Almost any is better than no plan at all.

The Countryside

Once again we must assume that policies have been devised and accepted for cleaning up the grosser pollutions and destructions in man's agricultural habitat. Cleaner streams, lakes to swim in, the scourings of strip mining and gravel pits removed, the land restored—all these are an essential part of a decent environment in the countryside. Now we must look at farmland and settlements in agricultural districts from two other angles of vision—as places to live and work in and then as places of recreation and refreshment for the inexorably growing numbers of urban people.

As with cities, we can say that the most urgent priority is to put an end to poverty and deprivation. All too often under cover of picturesque cottages and shacks hidden in graceful woods, living conditions are as unsanitary, disease-prone, and destitute as in the worst city slums. These poor conditions are complicated in country districts by the continued and massive exodus of younger farmers from the land. The United States, Western Europe, Japan, and, increasingly, the centrally planned economies of Europe are moving toward the condition first experienced in Britain—of having less than 10 per cent of the population working on the farms. A million farmers have left American agriculture since 1950. In France, a hundred thousand are leaving every year. It is clearly not desirable, in the interest of ecological variety, regional balance, and human quality that they should all be absorbed into shapeless sprawls of megalopolitan areas.

This is simply one aspect of a wider need—to determine how much rural life and agricultural output should be supported and encouraged at a time when technological and economic forces look like thinning out drastically both the variety and the human resources of country life. One objective in a wider policy under consideration in France is to build up regional economic centers—*pôles de croissance*—which in turn are linked to smaller communities for which they supply stabilizing employment, thus lessening rural poverty and stemming a little the migrations to the big city. The diversion of some industrial development from north to south in Italy is beginning, after two decades, to have its effect. It is significant that in the United States in the last ten years half the new jobs created in manufacturing have not, in fact, been centered on the metropolitan areas. There is an encourageable trend to more decentralized economic patterns, which in centrally planned economies can be achieved, as in Romania or Bulgaria, by purposive policies of directed industrial location and regional planning. These patterns are, incidentally, of critical importance for the even more urgent and different dilemmas of developing lands.

Yet as the number of farmers shrink and food is produced on smaller acreages, we confront a new environmental problem—the role of land that has been farmed and is so no longer. Some estimates suggest that in the next decade, something like 70 million acres of land that is now cropped in the United States may be withdrawn from farming. Parts of Western Europe are experiencing the same phenomenon—the lapse of one use with no very clear idea of the use that may follow. The point is that, in large parts of the country environment which delights and refreshes urban man, the maintenance, the work, the ordered detail, the almost parklike appearance are not provided unassisted. They are the work of the dedicated farmer. Remove him and the land that was the nation's pride and beauty will quickly deteriorate. Nowhere is this miserable aspect of disturbed and disordered land more lamentable than in the environs of spreading conurbations, where, to all the ugliness of nature drifting into weedy decline, are added the wastes, the gutted machines, the rotting hen coops of the retreating farmer.

In time, provided urban pollution does not drench such land with a fine, continual rain of dust and acids, nature will begin the task of rehabilitation. In the United States, all through the suburban areas of New York State and Connecticut, for instance, the old stone walls of vanished farms climb the rocky hillsides under the thick cover of a new generation of woodlands. Trees are the great healers of nature. Given a minimum top soil, a minimum rime of earth upon the face of the rock, they or their companions or competitors—shrubs and ferns and grass and moss—will return to cover the earth.

But if man applies his ecological understanding and skills to this natural process, it can be hastened and diversified. Dedicated local conservationists could take the lead in using woods, copses, and tree lines to hold the soil's fertility, prevent soil erosion, lessen the risk from wind-pollinated weeds, and enhance the sense of order and variety on the fringes of our vast areas of urban settlement. In doing so, they would at least lessen the impact of one of the worst aspects of uncoordinated growth into the countryside—the spreading network of roads, railroads, power lines, water-supply systems, sewers, telecommunications which split the landscape into pieces, slice the fields and forests, spoiling them aesthetically, decreasing their productivity, and upsetting the natural patterns of life. It may take years to overcome this uncoordinated slashing of the countryside, but it should not be tolerated in future expansion.

Needless to say, such a policy demands a firm control over future land use. In the landscapes of pretechnological man, one has again and again the sense that the city is the beleaguered unit. Its walls, its moats, its towers are built not only to stave off neighboring predators—barons or "masterless men." It is also the surrounding worlds of forest and swamps and wild boars that are being held at bay. The city and the garden—the enclosed areas where man felt he could reign—were to be protected from untamed, incalculable nature.

Today the reverse begins to be true. It is nature that must be protected against the cities, the smelters, the bulldozers, the automobiles of technological man. Unless the defenses are raised as land leaves the active maintenance of the farm economy, it slips over into the urban sprawl or stays in a derelict limbo between use and disuse or is invaded by wholly disturbing and inappropriate buildings like a mock Gothic villa on a wild seashore, a massive, all concrete, four-story hotel beneath the Attic grace of the Temple at Sunion, or the glass-and-steel duplex adding meretricious glitter to the sober beauty of a forested hill. In fact, such horrors can be perpetuated in active farming areas. But the risks are greater as acreages become more concentrated, marginal lands are given up, and small farms deserted.

However, several countries have shown that stout defenses can be raised. In Britain where, obviously, land is a limited and most precious resource, a Countryside Commission was created for England and Wales and a separate Commission for Scotland in 1968. It is under the overall authority of the Ministry of Local Government and Development (which is itself part of the overall Department of Environment) and has the right to issue grants and oversee the preservation of all aspects of country life for the citizen's use. This Commission can designate Areas of Outstanding Natural Beauty and can give local authorities up to 75 per cent of the cost of planting trees, maintaining woodlands, and employing wardens. There are more than twenty-five of these areas and they include such incomparable landscapes as the Sussex Downs, the Cotswolds, Dedham Vale (where Constable loved to paint), and long stretches of the coastline. Recently, the French government has designated rural areas of high cultural value as national parks, where the work of a thousand years of careful farming will be preserved, as Charles Péguy said of La Beauce, as "an endless reservoir for all the ages to come."

Preservation is the first problem. But there are also difficulties in arranging for the right kinds of access. Sometimes in their eagerness to improve the city weekender's access to nonurban beauties, local authorities wreck what people come to see by widening roads, straightening corners, felling trees, and even shaving, in a well-meaning way, the wild beauties of roadside and hedgerow to uniform suburban grass.

In the deep country, we meet a further problem. If everyone goes to the same beauty spot, what beauty will there be left to see? A partial answer is being tried in Britain. Tree-screened car-parks are being placed along the Pennine Hills, where country trails and bridle paths begin. Thereafter, access is only on foot. The trails through Germany's Black Forest or the Bruce Trail on the Niagara Escarpment in Canada allow walkers the enormous rewards of going so slowly that not a leaf, not a butterfly, not a singing linnet, or moss-covered stone need escape the eye. But bulldoze in the big roads, extend and pave the camps for the caravans, lay on electricity, build barbecue pits and in a year the linnets are gone, the stones are bare, the leaves grimy, nature's variety bleakly diminished by man's present inability to move unless accompanied by an incredible clutter of objects. There are, of course, places for concentrated campsites. But they should be carefully situated so that they do not destroy the landscape the campers came, in part, to see.

More and more countries in Europe, with an eye to the tourist trade, are beginning to think in terms of preserving their rural heritage and designating areas of greatest beauty. But perhaps modern urban man needs even more desperately what one might call a policy for Areas of Least Landscape Value. It is at the points where urban sprawls peter out or approach each other, where farms have given ground to the rough grass and weeds, where the rusting skeletons of cars lie in abandoned dumps that positive action is needed to end the steady diminution of the citizen's amenities, or, worse still, of their power even to recognize an amenity when they see it.

Yet a careful, scrupulous land policy, designed to bring back color and design and the use of trees and gardens, can work miracles—as it is doing, for instance, in Holland's Randstad. Derelict land can be planted with quick-growing trees, industrial areas screened with tall poplars and massed evergreen hedges. Boundary lines between urban "villages" within the conurbations can be given greater definition by landscaping, tree-growing, and the careful preserving of open space. And since this kind of detailed and dedicated reworking of the texture of the land can be done with least cost and most enthusiasm by citizens themselves, local amenity societies have an opportunity here to mobilize voluntary work or—as in many of the states of America—to back local bond issues which provide the economic and physical means of what could become a joyous "spring cleaning" on a national scale.

The Wilderness

We must repeat. One of the chief difficulties citizens confront when they go off to seek refreshment from unspoiled nature is the number of other citizens who are doing the same thing. With international tourism tripling in ten years, with the surge of visitors rising steadily in national parks and in all well-known areas of great beauty, the visions—of forests, of wild shore and open water—are obscured in milling masses of people trampling the turf, parking the cars, shooting across the virgin snow in snowmobiles, and braining the innocent swimmer with their passing speedboats. And once such invasions begin, once hot-dog stands assemble round every Walden Pond, where can men find nature in her primal state?

Should we want to? Is there not something unattractively limited and elitist about an environmental attitude which puts the solitude of the few above the enjoyments of the many? Is there not some danger in concentrating so passionately on the fate of the bald eagle that we have no time for a day in the country for the ghetto child? May we not earn the scorn Tom Paine felt for Edmund Burke when he wrote: "He pities the plumage and forgets the dying bird"?

But there are a number of sound, entirely unelitist arguments for the preservation of the wilderness and of the wildlife it contains. In the first place, there are a large and growing number of people who want to spend some of their time away from the pressures of the man-made order. There is enough wilderness in the world to increase very greatly the number of national parks and to see to it that some of them are preserved in their original condition with access kept so strenuous that the solitary walker is most unlikely to be crowded out. Of the 97 million hectares listed as national parks or reserves—the equivalent of less than 1 per cent of the world's surface—35 per cent are in North America alone, another 15 per cent in Africa. There is clearly room here for expansion.

Moreover, the lopsidedness means that, apart from the value of these places for recreation, they do not really fulfill their potential contribution to man's scientific and aesthetic interests. Preservation and conservation are not merely matters of catering to minority tastes. The still-untouched domains of nature, the still-living multitude of natural species are essential for the work both of the scientist and of the artist. They are needed to complete our still-patchy knowledge of the interdependence of living things and the underlying balances of the natural order not yet disturbed by man. They are needed to preserve the images of variety in plant and animal without which the human imagination could easily become a starveling. The animals, the plants, the biomes are entirely unrepeatable. Yet thousands of different animal species are already known to have been wiped out. In our own day, rapacious overkilling threatens to wipe out most of the major stocks of whales.

The lesson is obvious. How many essential species have vanished before man discovered how useful they really were? The whole concept of conservation—of studying balances and cycles, of habitats and species, of keeping seeds of all kinds in "genetic banks" is, in essence, an attempt to secure for man a "fall-back" position in case his overweening confidence or overwhelming numbers unleash on him unmanageable threats to his sophisticated hybrids, his extensive monocultures, and his urban deserts.

And this perhaps is the ultimate meaning of the wilderness and its preservation—to remind an increasingly urbanized humanity of the delicacy and vulnerability of all the living species—of tree and plant, of animal and insect—with which man has to share his shrinking planet. As he learns to observe their interdependence and their fragility, their variety and their complexity, he may remember that he, too, is a part of this single web and that if he breaks down too thoroughly the biological rhythms and needs of the natural universe, he may find he has destroyed the ultimate source of his own being. This may be too hard a lesson for him to learn anywhere. Least of all is it likely to reach him amid the perpetual thrust and din of his own settlements and inventions. But if somewhere in his community he leaves a place for silence, he may find the wilderness a great teacher of the kind of planetary modesty man most needs if his human order is to survive.

9 THE BALANCE OF RESOURCES

Problems of Calculation

NOWHERE DO CALCULATIONS OF COST, reflecting scarcity or abundance —price signals in market economies, estimated social costs in planned economies—confront greater complexity than in the question whether, when all the additions are made—of reserves, of materials, of pressures of rising consumption and rising population, of energy resources—there is or is not enough to go round to meet all the human demands that seem to be surging up on planet Earth. In fact, before any rational answer can be given, we have to make some kind of estimate of three factors—the likely scale of increases in population, the possible weight of their combined consumption, and the resources—of materials and energy—that may or may not be available. The difficulty of the calculation lies in the degree to which, in good ecological fashion, the three issues are wholly interdependent. The use of materials depends upon the numbers and expectations of the people using them, the scarcity of resources helps to determine their cost but also, by making them less attractive components of the industrial process, induces alternative and less wasteful uses. Our answers are full of variables which help to influence and transform each other. For the forecasters, life is full of risk. Their activity is rather like the game of croquet in *Alice in Wonderland*. The mallet is a flamingo that raises its head, the ball is a hedgehog that simply walks away.

Population in the Developed Lands

There are about one billion people in the developed world today— North America, Europe, the U.S.S.R., Japan, and Australasia. Their

numbers are not growing spectacularly fast. The two largest nations—
the United States and the Soviet Union—grow by about 1 per cent a year.
In other words, births exceed deaths every year by 1 per cent of the
existing population—by 2 million in America, by 2.2 million in Russia.
They are at the top of the growth league in developed states with Britain
at 0.5 per cent and Austria at 0.4 per cent at the bottom. But although
these levels look modest indeed compared with the average 2.5 per cent
and occasional spectacular 3.4 per cent rates in developing countries,
they have disturbing consequences.

The first is the scale of population density from which future growth
sets out. The United Kingdom (England, Scotland, Wales, and Northern
Ireland) may be only increasing by 0.5 per cent a year. But there are
already 56 million people in the country, over 80 per cent of them in
urban areas. This gives a density of about 147 per square mile of land
from which must be subtracted the high hills and bleak uplands and
much of Wales and the north. So if the increase in numbers is about 12
million (the official estimate for the next two or three decades), it can
exercise formidable pressure, given the islands' essentially limited space.

Or take another complication. Some developed countries, notably the
United States and Holland, have experienced relatively rapid population
expansion recently. The United States took nearly a hundred and fifty
years—and immigration which averaged a million a year in the early
1900s—to reach its first 100 million. The next 100 million arrived in about
fifty years, thus providing a much larger number of potentially childbear-
ing couples for the decades ahead. If they decide to produce an average
of three children, it will take only twenty-five years to add another 100
million, only ten for the next. This is simply a particular illustration of
the general fact that population growth, unchecked by disease, famine,
or war, feeds upon itself, each generation of mothers producing more
potential mothers for the next. It is this inexorable mathematics which,
all other things being equal, accounts for the fact that a 1 per cent rate
of increase will double a population in seventy years, a 2 per cent increase
in thirty-five years, or 3 per cent increase in only twenty-three years. And
because each time, the reproductive base is wider, the billions, on a world
scale, come faster and faster. The world's present average 2 per cent
growth will give us 7 billion by about the year 2000 and by then the extra
billion will be arriving in less than seven years.

To return to the American example, even if parents decided to produce, on the average, two children, some families with more, some with one or none, recent estimates suggest that America's postwar surge of growth would still keep American population on the increase well into the twenty-first century and a stable population of perhaps 370 to 400 million would hardly occur until the end of it. But an average three-child family would provide 300 million by 1989, 400 million by 2014, and nearly 1000 million by 2070. This is what demographic arithmetic makes of such apparently small adjustments as the difference between a two- and a three-child family—both, by earlier standards, of modest size.

And the concept of an added baby—minute, easily covered, lying in its cradle, and nursed by its mother—gives a totally inadequate picture of the impact on society, economy, and environment that the children of the developed world are destined to have. If in the calculations that follow we take an average American child, the reason is simply that America's material living standards are the highest in the world and that we have not too much reason for supposing that they are *not* the standards other families would aim at, if the "revolution of rising expectations" could run completely free.

Take first of all basic social needs, many of them provided by public funds. In many developed countries, they are not yet adequate. Education, health, housing, especially for the poorer members of society, could all be greatly improved. Even to take better care of the population growth entailed by a two-child family, public spending must increase. This is simply because the higher the density of population, the more decisions and expenditures have to be made collectively. Families hacking their way through the virgin forest had better be ruggedly independent. It is the only path of survival. But dense metropolitan populations have to solve many of their problems by collective acts and political decisions. The more people, the more they are concentrated in urban areas, the higher the public budget becomes. And this pressure is quite independent of the need for kinds of environmental expenditure not undertaken in the past.

If a society then adds an extra child to the average family, the collective budget must increase. Economically, this can certainly be afforded in developed countries, even though the number of young participants who consume and do not yet produce goes sharply up.

But if society as a whole—as is the case in a number of countries—shows political resistance to higher public expenditures, it is possible that more children will be catered for by a relative decline, per capita, in public spending on basic social needs and by a further postponement of environmental expenditure felt to be "less essential." Either way the overall standards of amenity in society are lessened.

If, on the contrary, citizens are rational and generous enough to understand that society as a whole suffers if its children are deprived and its environmental needs shortchanged, then public spending will make steadily larger claims on the biosphere's resources. In wealthy societies, with rapid growth rates, these claims need not be met by cutting back private claims. They are likely, by more rapid economic growth, to be satisfied concurrently. There is, after all, little reason to suppose at present that increased public costs will lessen private pressures for rising consumption. The final result can only be a further engrossment by a particular society of a still higher percentage of the planet's resources.

To return to our American baby, every child born into the American economy—taking the 1968 figures—contrives by the time he grows up to consume every year over a million calories and 13 metric tons of coal equivalent (or 2700 gallons of gasoline) in energy. He has probably nearly 10 metric tons of steel attached to him for various uses, particularly in the motorcar which is on the way to being owned by one in two of all citizens. He probably has another 150 kilograms of both copper and lead and 100 kilograms of aluminum and of zinc in use in his various appliances and artifacts. To keep him supplied with all these needs, the country's roads, railways, and freight planes transport 15,000 tons of materials per kilometer and to his door they bring the TV sets, the washing machine, the refrigerator owned by over 70 per cent of the population. They also deliver the second car and color TV to 30 per cent of the people and—a booming, expanding market—air conditioners to 20 per cent. Multiply this level of material abundance first by another hundred million citizens and then by the still "unfelt wants" implied in the concept of expanding consumption, increase it again by a jump to American standards by Europeans and Japanese, who in the critical area of energy use are, in the main, still more than 40 to 50 per cent behind —multiply *their* demand by the doubling of population, largely in cities, inherent in even a 0.5 per cent growth of population over the next half-century, and the calculus of food eaten, energy used, metals consumed, and transport systems enlarged suggests at least a tripling of present claims of developed nations on the planet's nonrenewable and exhaustible resources. And such an increase would be added to levels of consumption which already divert about 75 per cent of the world's nonrenewable resources to developed societies even though they contain less than 33 per cent of the world's peoples. The American baby with whom we started this calculus is in fact only a two-hundred millionth part of 6 per cent of the world's peoples. But he will help his community to create over 30 per cent of the world's drain on nonrenewable resources.

If we are primarily concerned about the problems of pollution, environment, and pressure on the world's resources, we have to reckon with the fact that the baby born in a more affluent sector of San Francisco, London, Stockholm, or Moscow will use a much larger proportion of the planet's resources than an infant appearing in India or Chad or Outer Mongolia. We know that to stabilize the world's population is at some point a condition of survival. The resources of the biosphere do not make up an unlimited system whereas the geometrical progression of reproduction seems to be so until famine and death massively intervene. If we prefer, as a species, to employ somewhat less brutal and indiscriminate methods of birth control, then we have to relate population to what the planet can support. It therefore follows that an American baby, who will require a million calories of food and thirteen tons of coal a year during an average lifetime of sixty-five years, is going to run through the biosphere's available supplies at least five hundred times faster than an Indian baby looking forward to fifty years with an annual consumption of perhaps half a million calories and almost no energy save what he will himself produce from those calories. The sheer increase in the numbers of Indian babies does, of course, change the calculus. But if our concern is to take unsupportable strains off the biosphere, the goals underlying population policy must include, for high-consumption societies, a family size at which their populations become stable.

This is certainly not an impossible aim. The invariable consequence of modernization with its increases in prosperity, wider education, and greater variety of work for women has been a rapid fall in the birth rate. If one subtracts from the births which actually occur the ones which parents would have wished to avoid, the fall could be even sharper. Once again, the United States provides some significant evidence. The Natural Fertility Study in 1965 suggested that if unwanted births were eliminated, the nation would be moving significantly toward a stable population. Its large sample survey of parents showed that perhaps 20 per cent of all pregnancies were unintended.

More difficult questions of public policy arise if the general unforced consensus of parents does not move in the direction of smaller families. The notion that tax and educational policies should be framed to discourage more than two children is attractive to many people but will be of doubtful benefit either to society or family if parental improvidence is taken out on unoffending children. Yet the consequences of improvidence are ultimately inescapable, whatever mitigations the government may propose. No imaginable disposition of the planet's resources can give the 1000 million Americans who could arrive, via the three-child family, within a hundred years, conditions as spacious and promising as the standards enjoyed by three-quarters of America's two hundred million inhabitants today. Nor do other developed lands escape the dilemma. Something has to give—family size, standards of living, or the biosphere's survival. Of these alternatives, stable family size with modest affluence seems the most humane solution.

But even this definition, "modest affluence," raises questions. The present birth rates and patterns of fertility in developed societies make it fairly certain that by the year 2000 they will contain about one and a half billion people. All modernized societies conduct their economies on the basis of rising material standards for all citizens. Most of them have contrived to increase their economic base by 3 to 4 per cent a year, some by considerably more. This growth has allowed a large increase in public services such as health and education. Life expectancy at birth has increased. In many countries the proportion of young people securing advanced education is comparable to that for secondary education a generation or two ago. Enrollment in graduate schools has been soaring and the skills represented by advanced education are possibly *the* essential input in society's increasing productivity—its ability to make "more for less."

The other critical index of productivity is the use of energy. This, too, has risen steadily in developed societies from an average per capita figure of roughly 2.75 metric tons of coal equivalent—the number of thermal units produced by a ton of coal—in 1940 to 5 metric tons in 1970. This rate, projected onward to the year 2000, gives a level of per capita energy consumption of over 11 metric tons. And this may be on the low side, since, with renewed spurts of growth, one or other of the developed

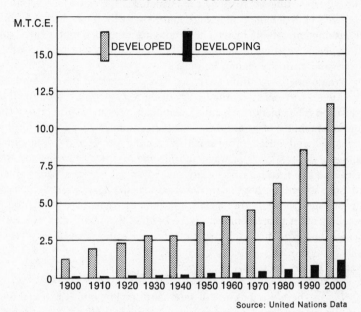

DEVELOPED AND DEVELOPING COUNTRIES
ENERGY CONSUMPTION PER CAPITA
IN METRIC TONS OF COAL EQUIVALENT

Source: United Nations Data

nations may catch up with the American per capita level of energy consumption. This, as we have seen, is already 13 metric tons of coal equivalent and is the result of a rate of expansion in energy use of between 3 and 4 per cent over the last decade. In our projections we may thus be underestimating the capacity of fully modernized, technologically adept, and profoundly educated societies to raise men's sights to new energy slaves, new goods, new services, new capacities and needs. It is this risk that makes our future figures so shaky. We take a confident jump onto the moving staircase and half way up it accelerates. Where are our projections then?

We can no doubt be sure that if the one and a half billion people in developed societies in the year 2000 had per capita incomes of, say,

$20,000 at 1970 prices and aspirations aiming at the $50,000-a-year level, the "load" of materials, in energy, in metals, in land use, in waste disposal—for which each would be responsible—would impinge in unpredictable and destructive ways upon the planet's underlying life-support systems. But where along the escalator is the exit to a level floor? Is there a kind of biological limit to man's desire for food, shelter, leisure, entertainment, talking by telephone, moving about in a motor car, flying in airplanes, visiting distant lands? Is there a threshold beyond which desire and curiosity cannot be pushed?

We do not know. But clearly we have not reached it yet. Standards are still set for society by the luxury of the very rich purveyed to all by television. The chief drive behind inflationary pressures springs from this upward mobility of aspirations. The pattern underlying a great deal of industrial, and indeed, professional bargaining is to keep the differentials between various levels of wage and income in order to make sure that if the poorer members of society are advanced, those ahead of them advance as well. Perhaps no modernized societies have reached a consensus on what a "good life" means in economic terms save in relation to reaching other people's higher levels. In all of them, any general state of benign satiety seems a long way off.

The Pressure on Resources

Let us therefore look not at internal discipline but at external constraints. In developed countries food does not look like being a problem this century. It is being provided by farmers who are retiring land, consolidating farms, and greatly raising output per acre. Even if some of their practices—with fertilizers and pesticides—put other sections of the biosphere under strain, the facts do not suggest that more careful and selective use would greatly reduce the output of food.

When we turn to the materials used in industry, we begin to encounter the extreme difficulty of deciding just how large a reserve of this or that mineral is likely to be available over what span of time. Take a critical substance like iron ore upon which man's industrial activities have been based for three or four thousand years. Annual consumption of iron ore has quadrupled since 1950, roughly 85 per cent of it in developed lands. If this rate of increase continues, estimates suggest that about 17 billion metric tons will have been used up by the year 2000—

at which point there could be only another 88 billion metric tons left, a calculation which might suggest exhaustion of the ore bodies by the middle of the twenty-first century.

But no such simple calculus can be made. These estimates were made on the basis of the 1968 price of $15 a ton. As reserves run down, four things will happen. Prices will rise, which will make it worthwhile using ore with a lower percentage of iron. This alone could triple and quadruple potential reserves. Next, a new wave of exploration will be unleashed which may, in developing lands, where prospecting has not been intensive, uncover unsuspected ore bodies with a high iron content or add to the scale of less valuable reserves. The third consequence will be much more careful use of scrap. For every ton of steel, about half is recycled metal, partly recaptured in the plant itself, partly rebought from scrap dealers. There has been, recently, a considerable growth in the efficiency of this use of scrap although some technologies tend to use less of it. In the design of new steel mills, recovery rates could certainly be improved still further and suitable incentives could encourage a higher proportion of scrap.

The fourth consequence will be a turn to other materials and to much greater research into reproducing in them the qualities that industrial processes require. Steel itself has been transformed over the last fifty years by the addition of a vast range of metals—vanadium, tungsten, molybdenum—which were once curiosities on Mendeleev's table and suddenly became the key to greater tensile strength or higher capacity to endure heat or better all-round plasticity. The plain truth is that iron ore at $20 a ton entails a completely different set of uses—and nonuses—than ore at $15. Nor is the calculus very different with other critical metals.

We can look at the copper equations. World consumption of copper has risen from 1.6 million metric tons in 1950 to nearly 7 million metric tons in 1970. If we extrapolate this trend of consumption at 1968 prices, another cumulative total of 400 million metric tons could be needed by the year 2000. But only 300 million tons are worth extracting at that price. In spite of a severe price fall recently as a result of the American recession and uncertainties over the nationalization of major copper industries in developing lands, it seems fairly certain that over the next decade, price changes will reflect growing scarcity and it will be worthwhile bringing ore bodies into production which have a much lower metal content. Copper is thus likely to become at once more available and more expensive. Further prospecting and more economical use would then probably extend its period of usefulness to man. The United States' present recycling of copper equals about 40 per cent of its production. The percentage would rise. Other materials would also be substituted for copper. Nonetheless copper seems to be on the planet's short-supply list and once its price begins to represent its real scarcity, other materials will be brought in as substitutes and lessen its use.

One of these substitutions is already taking place—both for steel and copper. Aluminum is the most abundant metal in the world and third in rank of the most abundant elements. Its consumption has nearly quintupled since the 2 million metric tons of 1950. When first used in the 1920s, it cost \$545 a pound. On occasion since, its price has fallen to as little as 15¢. It can be used easily in combination with other metals and can contribute hardness, lightness, durability, conductivity, and all the other characteristics of a good workhorse metal. As such it can be used in cars, in aircraft, in cables, in electrical and household appliances, in containers, mechanical limbs, venetian blinds, or modern sculpture.

Another ubiquitous element, silicon, makes up a quarter of the earth's crust. It is best known to us as sand and has long been used for glassmaking. About fifteen years ago, a chance overheating of glass in a laboratory experiment produced a substance that did not break but bounced. Analysis showed that the great heat had transformed the basic structure of the silicate molecules into a lattice or crystalline form resembling a metal which proved to have both enormous strength and a heat resistance of up to 1300°F. This was the beginning of ceramics tough enough to make the nose cones of rockets.

But the use of both aluminum and silicon reminds us of a wider fact. Once scientists could decipher the alphabet of the elements, they saw that the underlying components of all atoms—the elementary particles —are the same. It is the different arrangements and numbers of these particles—electrons, protons, neutrons—that give elements their varying characters and weights and determine how they link up with other atoms to form molecules. The linkages have a crystalline structure or lattice which can be studied by the relatively new techniques of crystallography, and once they are known, they can in theory and often in practice be imitated and manipulated to produce complexes with the desired durability, conductivity, or whatever characteristic that is being sought. Another method, not fully developed until the 1930s, is based on under-standing the complicated chains of carbon which make up the fundamen-tal cellulosic wall of plant cells throughout the vegetable world. With the help of high technology, scientists began to create artificial molecules as substitutes for silk, wool, cotton, for the long, thin, strung-out patterns of fibers, for the big, coiled chains of rubber.

In essence, it was a welding job, manipulating the molecules under heat, pressure, and the work of selected catalysts into a usable version of materials originally found in the earth's vegetation. It demonstrated the principle of substitutability over a field almost as wide as that of atomic structure itself. Plastics substitute a new basic kind of man-made material which, worked over and reinforced in a thousand ways—most recently by the addition of "whiskers" or minute filaments of boron, graphite, and beryllium—can combine astonishing strength and resis-tance with an equally astonishing diminution in weight. In some of the latest aircraft, for instance, hundreds of pounds have been saved by doubling the amount of plastics on the exterior skin of the plane. The percentage of plastics in automobiles is also steadily rising. And its all-purpose quality, combined with its lightness, makes it a substitute for a very wide range of metals. For instance, it has been estimated that in general one weight unit of plastics can replace nearly nine of steel, seven of copper, six of zinc, and three of aluminum.

It is this basic new ability to rearrange matter into a whole variety of forms and uses that makes it very difficult to calculate possible short-ages in the world's supply of particular substances. Equally, it makes estimates of the effect of such shortages almost impossible. Can we then conclude that we have really not too much cause for worry? Is the discovery of atomic structure and the alphabet of the elements really the philosopher's stone? By sufficiently pushing and prodding the lattice of lead, can we come up with bright gold? After all, graphite has been made into industrial diamonds. If the alchemists are right and everything can ultimately be made with anything, then "planet eating" may not be so dangerous. There is a great deal of it to eat.

The flaw in the argument lies in the complexity of the technology, the scale of the energy through which the transformations have to take place, and all the increased costs and environmental disruptions such transformations can entail. The greater costs in extracting 0.5 per cent ore from metal bodies, when formerly industry was using stocks containing 3 per cent, is due in part to massive strip mining, in part to more elaborate technology and much greater use of power. Aluminum, for instance, requires twenty times the energy to extract from ore than is the case with iron. A large petrochemical plant, producing gasoline, diesel oil, and a whole variety of basic raw materials for synthetic fibers and detergents uses up considerably more energy and infinitely higher technical skills for each unit of production than the older, more labor intensive industries producing raw coal, cotton goods, and soap.

Similarly, the cars which modern technology supplies with gasoline eat up far more energy than horses and carts and also require more inputs in road construction and repair than did the one-shot construction of railroads, which they have replaced in many fields. The astounding increase in the use of energy in the United States, which has put it 50 per cent ahead of any other developed nation, may be in part a passing phase connected with an unsustainable degree of automobile use and household gadgetry. But it may also represent the rising trend of modern technology. We may not be running out of materials. Are we running out of energy?

The Energy Equation

The answer to this question is fully as complicated as the answer for raw materials. There are many different kinds of energy. They are, in the main, interchangeable. If any one of them becomes scarce and its price begins to rise, there is a very great impetus given to technologies, which use it more sparingly or switch demand to other sources. There would be real shortage only if *all* sources were in danger of becoming exhausted and world demand were rocketing upward toward that point.

Demand is, of course, rising fast. Some estimates suggest a steady 4 to 5 per cent annual increase in total world energy demand until the year 2000. This requires doubling the supply every few decades. If we take the figures for electricity—by far the most rapidly growing form of energy —American forecasts suggest an increase of at least 300 to 400 per cent in the next twenty years. Other developed economies are growing even faster, although from a lower base. Japanese use of electricity, for instance, has been increasing by 15 per cent a year. All in all, we may conclude that technically it may be possible to meet such fantastic rates of expansion. But two things are not in doubt—a vast increase in capital costs and a massive impact upon the environment.

Let us look first at power resources which are inherently exhaustible because they are provided by the fossil fuels—coal, oil, and natural gas —which were stored in our planet during the geological upheavals of past billennia. This inherited "capital" of energy has carried the Industrial Revolution so far—first through coal, which provided at least 80 per cent of the world's energy as recently as 1920, and now by petroleum and natural gas, whose shares have driven coal down to about a quarter of all energy consumption.

Of these three, natural gas has the smallest reserves. Its freedom from pollution makes it popular now, and this advantage might keep it in use when increasing scarcity raises its price. But continued use would only hasten exhaustion.

Petroleum, too, is on the short-supply list. One of the most quoted estimates for usable reserves is some 2500 billion barrels. This sounds very large but the increase in demand foreseen over the next three decades makes it likely that peak production will have been reached by the year 2000. Thereafter it will decline. Supplies could also be affected by price changes. Oil-producing countries are already using their joint influence to raise that price. This would stimulate the conversion of shale to oil, and it is estimated that the Athabasca Tar Sands in Alberta, Canada, are equal to at least 300 billion barrels. Moreover, a higher price would hasten the introduction of techniques and processes for refining oil and gas from coal. But the sheer scale of the demand for petroleum —in transport, for power stations, for the production of petrochemicals of all kinds—almost certainly means that, by next century, it will *not* be available to power anything remotely like a world population of motor-cars at American densities of ownership. Nor is it likely to provide fuel for electricity generation. Its most valuable use looks like being the provision of materials for petrochemicals and these too will make their claim on dwindling reserves as prices move inexorably upward.

What will be available to take its place—in transport and electrical generation? Clearly coal looks like being with us for quite a time. World reserves are above 5 trillion tons and even the most pessimistic observers believe it can last us out another century. It will remain a major source of power for generating electricity and its use in electrically powered mass transit and even in nonpollutive cars is bound to increase.

There are, however, as we have already seen, high environmental costs involved in a big increase in the use of coal for generating electricity. New plants will have to be built which emit only limited amounts of soot and fly ash and sulfur oxides. Nor can they be allowed to increase thermal pollution above acceptable limits. Gassification of coal may, before too long, reduce some of these costs. But its rapid development will be speeded up if the price of electricity covers present pollution and puts an end to what is often a serious underpricing of electricity.

Moreover, wherever a new demand for sulfur-free coal leads, as it threatens to do in the western regions of the United States, to a massive increase in strip mining, the ultimate price of the power should cover the rehabilitation of the land. It should also include the cost of proceeding at less than bulldozing speed on the chance that other less pollutive techniques may emerge before parts of the planet's surface are made to look as barren and rugged as the moon landing above Hadley Rille.

What are our hopes for less polluting types of energy? The cleanest and safest are also among the most inexhaustible. After animal power, windmills and watermills are man's oldest sources of nonhuman energy and with so much wind and water in the world, it may seem odd that new technologies have not turned this vast potential income into usable supplies. Moreover, the sun itself, safely shielded from us by banks of oxygen and ozone, streams down day after day its inconceivable energies upon our planet. Is there no more direct way of plugging ourselves into these daily supplies of which we use only one-third of 1 per cent? If any such technological breakthrough proved possible, we could then look back upon man's rapid exhaustion of fossil fuels as simply the "self-starter" for his vast energy system which, invented by the technologies which the fossil fuels made possible, plugs the planet into cosmic supplies and carries it along at acceptable levels of self-renewing and inexhaustible energy.

The difficulty about the concept is that men have not yet found the way of directly tapping the energy cycles of the cosmos on a sufficient scale. There are still large opportunities for more hydroelectric power, derived from damming rivers and using their fall to power the turbines. But these sites are, in the main, in developing lands and offer some of them valuable nonpollutive sources of energy.

The harnessing of tides—which is being tried in La Rance in France and in the Soviet Union—seems also to be limited by the number of sites at which the rise and fall of water is on a sufficient scale and also by still very large uncertainties about cost.

Solar energy is not too safe an investment in the cloudy temperate regions where the demand for power is most urgent. At present levels of technology, it still requires a very large square mileage to produce an area within which enough incoming sunlight can be sufficiently concentrated to produce energy. However, in cloudless regions it clearly has a part to play and should be combined with research into its possible use in desalinization projects.

But the most immediate alternative to shrinking supplies of fossil fuels in any massive satisfaction of the world's future power needs lies in the various forms of atomic energy. And here we confront, with a seriousness which demands the utmost integrity of judgment and depth of human care, the profoundest implications of the Promethean legend.

Promethean Fire

What we are doing, in basing a larger and larger part of our energy supplies upon atomic energy, is bringing down to earth the powers which would have never permitted any kind of organic life to develop on this planet, had not billennia been spent in building up protective mechanisms—the oceans, the first creations of oxygen and ozone, the breathing out of the all-encompassing atmosphere by the earth's growing cover of green plants.

To generate this power on earth is, almost literally, the Promethean act of stealing the fire from the gods. And we should remember with an uneasiness born of a sense of sacrilege that the first use of this fire was to wipe out two cities, without care for individual guilt or innocence, man or woman, youth or age. In fact, if the compulsions of war had not so vastly hastened atomic research and had there not been an immediate postwar concern about fossil fuels, it is possible that developed societies would not have embarked with such speed upon processes which are hazardous in quite new ways, above all because they create the risk of irreversible damage to mankind's genetic inheritance. The only sane attitude to the massive use of nuclear energy, to predictions that it will account for a quarter of the earth's energy by the end of the century, is therefore one of the utmost responsibility and the most extreme caution.

What are the prospects for the peaceful uses of nuclear power? One type is now available, two others are in the planning stage. The present generation of atomic plants uses the same processes as the first atomic bomb—energy released by fission. The unstable uranium isotope 235 makes up less than 1 per cent of uranium, which is largely composed of the stable 238. This small percentage has to be separated out and the result, enriched uranium, is placed in a core in a nuclear reactor. It is then bombarded by neutrons, thus setting in motion a chain reaction. This process also produces new combinations of atoms and one of these elements, plutonium, is also unstable. Like U-235, it breaks up under bombardment and releases more neutrons. As they all bash about and collide, they also release some of the U-235 which exists, in minute percentages, in the uranium blanket. The risk that the whole chain reaction will end in an uncontrolled release of radioactivity is offset by inserting rods of spongy absorbent elements—cadmium, boron, hafnium —into the core and this slows down the reaction sufficiently to produce energy which then heats steam as in traditional electricity generators.

The limitation on this process is the scarcity of U-235. Uranium oxide still costs no more than $15 a kilogram and at this price the nuclear plants are competitive with coal-burning plants—provided we ignore such huge subsidies as the original development of nuclear technology by govern-ments for war and the element of subsidies in America's practice of providing insurance cover of the order of $560 million to each nuclear plant. But if more and more atomic stations are to be built—there are more than a hundred planned and Britain already depends for 13 per cent of its energy on nuclear power—the pressure on supplies of scarce uranium 235 may alter the cost calculus and either slow down the kind of expansion predicted at present or make more supplies available, but at a higher price.

But two other approaches to nuclear technology virtually by-pass the problem of fuel supplies—the breeder reactor and the fusion reactor. If the present experiments in breeder reactors are successful, the neutron power released by bombarding U-235 will go on to transform the whole mass of U-238 into fissionable plutonium as well. Instead of a 1 per cent use of uranium oxides, the plants will move on to a 70 to 80 per cent exploitation rate, actually producing more fuel than the original input needed to set the chain reaction into motion. This will vastly increase the usability of uranium reserves and introduce unpredictable changes into the structure of prices.

At this point, the economies of nuclear power become transformed in ways that it is still difficult even to grasp. If all uranium came to be fissionable, the oceans, which contain 0.3 milligrams of uranium per cubic meter, or granite, which contains 0.03 per cent, could theoretically provide unlimited supplies. Even if the price were relatively high, fuel is not the major determinant in the cost of nuclear power. It is the installations and safeguards that largely determine expense. The chief problem with nuclear fuel is more availability than cost. And the oceans are not the only source of potentially fissionable material. The 1250 square miles of Chattanooga shale which occupy about 2 per cent of the area of Tennessee contain .006 per cent uranium. In terms of fuel for breeder reactors, 17 square miles would be equivalent to the whole present coal reserves of the United States. Three square miles would be the equivalent of America's entire petroleum reserve. The reserves as a whole would equal all the fossil-fuel reserves—coal, oil, natural gas—available to the United States since its origins.

It is these fantastic potential availabilities of fuel for breeder reactors —if they are successfully developed—that somewhat lessen the economic interest of the third possibility—the production of energy by nuclear fusion. This is, literally, the transfer to earth of the sun's processes, the vast release of energy achieved by the fusing of nuclei within the superheated hydrogen plasma of the sun's core. The amount of fuel available in the world for such a process is virtually unlimited—the hydrogen isotope, deuterium, in the oceans is the most likely eventual source. Fusion technologies also present the possibility of transforming the released nuclear energy directly into usable electricity. The stream of particles—electrons, protons, helium ions—released from the plasma by accelerated neutrons bombarding the deuterium would pass through a series of magnetic fields at decelerating speeds, then electrons would be diverted and positive ions collected in such a way as to pass into direct current without going through an intermediate steam-powered generator.

But the technical difficulties still to be overcome are enormous. The plasma has to be heated to a critical temperature before nuclei are released—the lowest successful one so far is 72 million degrees Fahrenheit. The plasma also has to be dense enough to produce a chain reaction. Otherwise the nuclei flash past each other without colliding and more energy is used up to get them moving than is later released. Plasma also has to be "contained" long enough in this superheated state to produce usable electricity. No physical container can withstand such heat, and magnetic fields seem at present the only possibility. The central difficulty is to achieve containment, density, and temperature all in a single system. Soviet and American physicists are working closely together on the problems and there seems to be some expectation of progress in the next two decades.

These are the systems. What are the advantages? In some respects, the present fission reactors, most of them light water-cooled systems, are less pollutant than unregenerate fossil-fuel stations still pouring out the old filth. Their costs are competitive. They are of obvious interest to countries without fossil fuels. Their future comparative costs depend upon getting beyond the narrow horizon of U-235 as the whole energy source. Otherwise, coal would seem to provide a more economical fuel for some time to come, even with the added costs of removing pollutants. The real breakthrough economically would be the breeder reactor which could power the world for another millennium and still leave fusion power in reserve.

What are the hazards? Both fission reactors and breeder reactors produce high thermal pollution. In this they offer no advantage over conventional plants. In fact, the discharges can be hotter. In contrast thermonuclear process would probably generate less thermal discharge since it would convert fusion directly into usable energy.

Fission and breeder reactors pose the problem of radioactivity—both in their effluents and in the risk, however small, of a major breakdown. A very large effort is made to contain all the radioactivity given off during nuclear generation within hermetically sealed containers. In fact, a significant part of the cost of nuclear construction consists of building maximum safeguards into the design. But some seepage occurs—in the cooling water, in released gases, in the cleansing of the fuel elements— which poison the reactor if they are not taken out and reprocessed from time to time. Seepage also occurs in the disposal of nuclear wastes.

Some of the radioactive elements released are very short-lived, a matter of a day or two. But some have a half-life of a hundred years, some of a thousand and more. As nuclear generating becomes more widespread, the safe disposal of by-products such as strontium 90 and cesium 137 will present ever-larger problems of storage. At present longer-lived by-products are treated first by dilution, then evaporation. The remainder is put in huge stainless steel tanks. In twenty years, there appears to have been no major leakage. But the whole atomic program is still on a relatively modest scale. Other methods under research include solidifying the radioactive wastes in glass or clay like flies in amber. In the United States the plan is to stow casks away in old salt mines. If, as is expected, installed nuclear capacity grows to between 100,000 and 800,-000 megawatts by the year 2000, the storage of liquid wastes and fission could be multiplied more than a hundredfold. While such a burden of materials is being retrieved, transported, and stored, the risks of leakage must obviously increase enormously.

There is, of course, constant monitoring and control all along the way. The amount of radiation any worker may receive in nuclear industry is minutely checked and since 1928 the International Commission for Radiological Protection has set increasingly stringent standards of safety. The fundamental difficulty, however, is that there is probably no safe threshold of radiation and that any dose can, under a wide variety of different conditions—age, exposure, metabolism—create genetic damage or cancer. The only safety rule that can be formulated is a "maximum permissible dose," representing the lowest risk compatible with nuclear activities deemed to be useful to individuals and to society.

When such an idea of "dosage" is translated into what may be preventable deaths from cancer, it has a callous ring. Take, for instance, the ICRP's current suggested standards for protection. The unit of measurement is the rad, corresponding to 100 ergs of radioactive absorption in one gram of tissue. For the population at large, the maximum individual exposure is fixed at two rads over thirty years, although it is admitted that such a level could result, for a population the size of the United States, in 2500 additional cases of cancer each year. Since everyone in his or her lifetime receives some radiation—from the sun, from x-rays in medical treatment—the question is whether emissions from nuclear power stations will push the level above the two-rad standard and thus intensify the risk of cancer and genetic damage.

And here the debate begins and has been engaged most vigorously in the United States. On the one hand, the Atomic Energy Commission recently reduced citizens' acceptable exposure to radiation coming from the present generation of light water-cooled nuclear reactors to less than 1 percent of previous federal standards. This corresponds to 5 per cent or less of citizens' exposure to natural radiation from the sun's cosmic rays. Moreover, nearly all America's reactors have, it is claimed, been operating within these limits for some time.

On the other hand, it is argued by critics that the change in standards occurred only *after* a strong attack upon atomic policy had been made by a group of scientists who claimed that the ICRP's standards underestimate the risk of cancer by a factor of at least 12. It is their argument that a vast expansion in nuclear production would, inevitably, increase overall exposure to radiation. Some critics have gone further and have stated that no expansion can justify a demonstrable increase in cancer.

Thus we come to the fundamental question. Is the plunge into nuclear energy worth the danger? That citizens do regard death as a reasonable risk in return for some supposed good is quite clear. Otherwise the 50,000 deaths a year caused by the automobile in America would have long since lessened its attraction. On the contrary, the number of cars continues to go up. The relative steadiness in cigarette smoking in spite of its direct links with lung cancer, emphysema, and heart trouble is another instance of risks being taken for the sake of desired ends. Citizens would obviously prefer their energy to come unburdened with any risk —not the death of a single miner or bronchitic sufferer or hotline maintenance man. But if the choice is, as it may be by the beginning of the next century, between insufficient energy and a relatively small increase in cancer and genetic hazard, then, on present evidence, the citizen will prefer the energy. He may do so with all the more readiness if he consoles himself with the fact that cancer estimates are based on, in the main, extrapolations to man of the results of experiments on mice. Such extrapolations give at best uncertain evidence upon which to base a rational choice.

Yet the citizen must be concerned with more than the insidious seepages of radioactivity which occur even in medical uses. How about the more sensational risks? No large nuclear reactor has yet produced an uncontrollable chain reaction but as the number increases, the chances of one doing so inevitably rise. The layman may believe and hope that all possible precautions and warning devices have been incorporated in the reactor's design. But he is not reassured when responsible scientists tell him—as has recently occurred in the United States—that there is no "fall-back" position, if, for some reason, the emergency water-cooling apparatus, upon which safety in present water-cooled reactors is based, should fail to operate. The nuclear core might melt and produce massive release of radioactivity into the earth equal to that from an atomic explosion. The criticism is based upon an alleged lack of testing and experiment in this vital area of safety and a demand has been made that no more licenses be issued for the construction of further nuclear reactors until the whole system of safeguards has been thoroughly overhauled.

In all this debate, one thing is certain. The ordinary citizen cannot judge the scientific facts. What he can and must do is bring his reason and common sense to bear on his country's whole approach to the problem. The first act of sanity is to insist, with all possible urgency and influence, on the need for caution. It cannot be said too often that we are now playing about with the primal energy of the universe. Any carelessness, any casualness, any calculus based upon inward-looking national advantage (and prestige) or on a quick profit gained by some smartly turned commercial deal is utterly unthinkable in this context. Men are not making a simple calculus of gain or convenience. They are confronting their own survival and that of their children and grandchildren and the whole race of man.

Caution in general is the beginning of wisdom. It also points in a number of specific directions. No country, whatever its system, must permit standards of safety in nuclear activities to be set by those most interested in increasing the use of atomic energy. The men and women concerned may be entirely disinterested. But there is no way in which the citizen can be sure of this unless the functions of safety and inspection are separate from promotion and production. In fact, the whole record of inspection in commercial societies shows that without this separation, interest and objectivity cannot be kept apart.

Equally, there is immediate need to end a commercial free-for-all in the international sale of nuclear reactors. Once let invention and production be determined by a competitive national search for prestige and profit—as it is today in the race for the first commercial breeder reactor—and the temptation to cut corners, lower safety costs, gamble on a breakthrough will become almost irresistibly strong. We simply cannot apply the liberty of unregulated competition we allow in the sale of swimsuits and hair curlers to machines and systems which may destroy human beings and put future generations at risk. No doubt the cry will go up that any form of public control over nuclear development and sales will disastrously slow down research and progress. But this is not a disaster. The availability of fossil fuels—especially coal—gives us a breathing space. At the very least, an international inspectorate of unimpeachable scientific integrity must be the licensing body for reactor sales.

The citizen, too, may apply some common-sense rules to more detailed issues. He can protest at any attempt to build nuclear stations near big centers of population. What if it does add to the costs of distribution? No cost would equal a nuclear leak in an urban region with lethal radioactivity spreading out for as much as seventy-five, miles in every direction. He can probably make a rational calculation that it is very much worthwhile to put nuclear stations underground, each of them linked with the plant for reprocessing fuels, the whole system connected by undersurface communications and possibly located near a convenient site for storing the wastes. If the construction costs add to the price of the power, it will be a true price and not one that treats human survival as a free good.

The citizen can also ask that, on the basis of entirely independent expert opinion, the monitoring systems in plants and over wider areas are set at the lowest possible contamination rate compatible with the continuance of nuclear technology. Since there are special risks involved in reprocessing fuels, transporting them, and disposing of wastes, every effort of inspection and research should be undertaken to see by what changes in technology and increases in precautions such risks can be met. And if this increases the costs of atomic plants, once again, these are reasonable costs that must be borne. We have already made the error of regarding air and water as free goods and have thus indirectly subsidized the cost of power and hence its bounding upward curve of use. If we are to treat the risk of rising cancer and genetic mutations as another free good, then we may get more power and a degenerating population. The costs of the highest levels of safety have to be carried. In any case, they look like being small compared with military insanities of twentyfold nuclear overkills and all the atomic gadgetry of a holocaust from which a recognizably human species will hardly survive.

Should the citizen go further and simply say that he prefers to do without the Promethean fire? That nuclear energy is simply too risky and too expensive?

Let us look first at the question of risk. The difficulty is that it may well be a choice of risks. Against the damage that nuclear energy may kill and maim human beings, we have to set other hazards. There are already three and a half billion people on the planet and there probably will be about four billion more by the year 2000. To feed them, clothe them, and shelter them, more energy will be needed on a scale that fossil fuel cannot satisfy for very much longer. Already, a percentage of those born are maimed by lack of protein in the first year of life; they die not from cancer but from malnutrition combined with all the intestinal diseases of dire poverty.

If the world population which is already on the way is to be better fed and housed, if the ten to fifteen billion who will appear before populations stabilize are to have anything like the indispensable minimum of consumption, new sources of energy must be found and the only technology visible on a sufficient scale at this moment is atomic energy. Even if citizens in already developed societies decided to check the rise in their own energy demands—which is not inconceivable—and even if they lessened them—which is harder to imagine—the sheer basic needs of all the world's peoples could not be met by rationing the energy of the already rich. Nor could the larger crops be raised, the new schools and clinics built, the citizens rehoused without that extra ability "to do work" that energy represents.

It would therefore seem rational to take every precaution with nuclear energy derived from fission reaction, to slow down precipitate deployment of untested technologies, to devote the possible breathing space provided by fossil fuels to more research, and to further work on every kind of alternative—hydroelectric where the sites are available, solar energy, fuel cells, geothermal tapping of the earth's internal heat, thermonuclear fusion. But to keep seven to ten billion people alive and reasonably well served on this planet, atomic energy looks like being the most likely answer. The alternative—of too little energy—would cause infinitely larger rates of malformation and death.

Who Bears the Cost?

It is, however, true that the more careful we are over atomic energy —as we must be—the more expensive at least in the short run, energy is likely to be. And this is part of a wider economic problem. We can be certain that the kind of public expenditure needed to end the worst pollutions of urban poverty, to install the kinds of treatment plants require for the disposal of wastes and effluents—particularly radioactive ones—to upgrade disturbed and ruined landscapes, to acquire public parks and recreation areas must entail a sharp increase in spending, mainly from taxation. In centrally planned economies, this increase presents fewer difficulties than in market economies with various mixes of government participation and intervention. But in any economy, the dedication of more funds to environmental needs in the broadest sense must mean some increased restraints on personal consumption goods.

We have very few calculations of what a full-scale environmental cleanup would cost. Certainly, no valid figures exist for anything as remote as the year 2000. But in the United States, which once again we choose because material standards are highest and traditions of government involvement probably least rooted, a calculation has been made by a private organization, the National Urban Coalition, of what a number of environmental costs might be. In general terms, the Coalition estimates that a decent human environment in a predominantly urban America would require well over $100 billion in extra federal spending over the next half decade.

At the same time, the price of a range of consumer goods could rise as a result of a large-scale industrial effort to check pollution. In the United States, the Council on Environmental Quality estimates that over $3.5 billion was spent by industry in 1971 to clean up air and water. It is difficult to believe that such costs will not reappear in oil prices, electricity rates, and consumer goods generally. Car prices, too, will certainly rise as the result of newly installed antipollutant devices. All in all, any kind of big cleanup coming on top of existing inflationary pressure cannot fail to have unsettling and unpredictable effects on the economy.

This situation spurs different people to different responses. There are those who argue that the only way in which the problem of environmental renewal can be met is by slowing down a rate of economic growth which has in it a number of highly irrational elements. The root of the trouble, they argue, is the spewing out of ever more effluents from factories and power stations to produce a vast number of now basically superfluous goods. Often they have built-in obsolescence. Often, the argument continues, they replace good, solid, less-polluting technologies —soap instead of detergents, for instance, or synthetic yarns instead of cotton. Under the usual kind of cost accounting there are, furthermore, no indicators for true amenities, because leisure and personal welfare, benefits gained or forgone cannot be expressed in a monetary transaction.

If, the critics conclude, we could take the pressure off our environment by a less passionate pursuit of consumption goods, by less obsession with innovations and gadgets, by a more modest use of energy, our environmental problems would be lessened for us simply by a steady decline in our propensity to "spend and spend" and pollute and pollute.

At the other end of the spectrum are those who argue that the essence of the system is precisely innovation, the drive for competitive advantage, the hope of great gain to follow on success, the search for that "better mousetrap" for which mankind will beat a path to the inventor's door. Take away the incentive, the drive to achieve real productive gains, and there will be no rising curve of wealth with which to rebuild the cities, clean the airs and waters, and still give the ordinary citizen the consumer goods of which he at least still gives no sign of being weary. And if you cut down on the citizen's chosen pleasures and tell him out of a declining income to provide for $100 billion more of collective needs, there is really not much in contemporary politics to suggest that his answer will be to give you his grateful vote at the next election or to forbear from strikes and riots if they are his chief means of protest.

In short, whatever your system—collective, private, mixed—increased public expenditure means less private choice unless a society's productivity, its means of production, its real resources, and inventiveness are growing and public needs and private demands can both be met. So do we end in an impasse? Do we confront yet another treadmill situation—more economic growth needed to provide the resources to clear up the mess made by economic growth and in the process creating still more economic mess to clear up?

The answer can only be provisional. Clearly if we go on as we do now over the next thirty years, with a billion more people due to arrive in developed economies, which are growing by 3 or 4 per cent a year with inadequate controls on pollution, speculative land markets, built-in obsolescence, and untrammeled energy use, then we are heading for ecological disaster.

But none of these things is an essential part of a responsible and well-managed program of environmental awareness and economic growth. Once again, perhaps, people find themselves caught up in single-thrust solutions—no growth or all growth, zero population or no family constraints, no market or no planning, no hope or no problem. But if we model ourselves in this debate—as in most other human activities—on our ecological systems, we find, surely, that we do not achieve balance by any one line or solution but by a careful interweaving of a great variety of partial solutions which added together do not produce definitive answers—nature is too dynamic for anything so secure—but give us the possibility of proceeding without disaster, correcting, reconsidering, backtracking, advancing, observing, and inventing as we go.

After twenty-five years of hell-for-leather boom, it is right that the greater weight in thought, attention, and policy should lie on redressing the needless consequences of expansion. It is significant that in a recent and courageous public statement, the government of the most rapidly growing of all developed countries—that of Japan—expressed, in the most dramatic way, its sense of need for new priorities:

> Too eager to raise our living standard and too anxious to catch up with the material prosperity of Western nations, most Japanese leaders have been the captives of the disease of growthmanship, even after such attitudes became unwarranted. Businesses, with a few exceptions, have paid little attention to what their activities would do to environment. Scholars and journalists have not realized the deterioration of environment or, at any rate, have not tried hard enough to demand preventive measures. The government has spent too large a proportion of public funds for productive investment, neglecting social services. It has failed to make adequate zoning plans to segregate the residential from industrial areas. It has done little to regulate the activities of polluting businesses except when the harm was apparent.
>
> In the meantime, environmental disruption has proceeded in a creeping way without drawing much attention at first, but then, with the quickening pace of technical progress and urbanization in the postwar years, it has developed into a monstrosity.

The Japanese case is certainly exceptional with its postwar growth rates of rarely less than 7 per cent a year. Exceptional, too, is the forthright honesty of its government's analysis. But at more modest rates there is reason to believe that a sane expansion of human resources can be achieved without a "monstrosity" of environmental degradation.

First of all, let us remember that developed economies have a "slack" to take in of at least $150 billion devoted every year to the most senseless, wasteful, and inflationary activity, the production of armaments. All governments of genuinely "advanced" countries know that war between them will be nuclear and hence the ultimate and irreparable pollution. Most of them might now reflect on such lessons as Algeria or Vietnam and ask whether limited wars, involving great powers in less developed areas, can be anything but profoundly wasteful and counterproductive. The largest environmental benefaction any government could bestow on its own people is first to keep the peace and then reduce the cost of preserving it to the levels of an international police force. To this point we must return. Here it is mentioned only as a reminder of how much of present economic pressures spring from idiot preoccupations with overkill nuclear weapons and preventive arms.

Next we can remember that the kinds of innovation needed to deal with pollution and the better disposal of wastes are themselves potential growth industries and employment creators. The sales of equipment for checking air and water pollution have been going up by 15 to 20 per cent recently. This may seem a backhanded form of growth but it is no more irrational than our present calculus of gross national product in which the output of effluent-producing industries appears as a plus and so does government expenditure to clear up the mess. Besides, the technology of pollution abatement has hardly got off the drawing board. The whole argument for effluent charges or higher pollutant controls or greater nuclear safety is precisely that they will speed up the elaboration of techniques which are incorporated not as extraneous and expensive extras but as integral and hence cheaper parts of the original design. If the right pressures are kept up, there will be no major technologies by 1990 which treat air and water as free goods for the simple reason that they will no longer be so. Moreover, still on the side of production, many of the techniques used to clean up atmosphere and water and to dispose of solid waste will, as we have seen, actually recapture materials for use and enable a sustained rate of growth to be maintained with much less strain on the earth's resources.

Standards of Quality

Such changes on the side of industrial activity could be reinforced by a new look at the nature of consumption in the national accounts. Economists are at work on the construction of better indicators of welfare. Ten years from now, our concept of GNP may include little boys swimming in the Delaware or the Volga, days gained for industry which were formerly lost to bronchitis and head colds, the drapes and clothes no longer due for the cleaners, the number of days without smog in central cities, the vigor and serenity which comes from walking to work in friendly neighborhoods, the lessening in police and prison costs, the rise in leisure as people begin to like sitting beside their city sidewalks, to find their neighborhood parks alive with rock or Bach or the police brass band, to walk in countrysides and wildernesses that are cared for and protected.

This question of the quality of life is surely one that can be combined with the concept of growth provided we take a broad view of the goods for which we are prepared to pay. A consumer good is something a consumer wants. If he is still caught in the traps of poverty, bad housing, poor health, and no opportunity, he will not be grateful if you offer him clean air and water instead. Basic needs *must* be satisfied in any civilized community by whatever degree of income redistribution is needed to provide public funds without grave inflation. To do less is simply to neglect the worst of all pollutions—the pollution of bitter, hopeless, neglected poverty. There *is* an inevitable and essential element of a redistribution of resources underlying the problems of the environment, just as there is a fundamental issue of social justice underlying every political order.

But once basic needs are satisfied, the wants that come thereafter do not have to be solely cars, clothes, frozen food, or electric can openers. If citizens want the nonpolluting pleasures of great art, if, as in tragic Hamburg in the winter of 1945 they ask for the Opera House to be reopened before a single bombed-out house has been rebuilt, then the output of opera singers and ballerinas will go up and the money earned by industrial or office work will pass through the nation's box offices, not the nation's waste-disposal systems, to pay for the ballets and operas. If citizens want to parallel their working life with renewed efforts of self-education, the output of teachers will go up and fees for adult education will pass through the colleges to pay for them. If they want more sports, they can provide themselves with more quite unpolluting sports grounds. If they want better and more varied television, they can spend more money on it directly and not receive it as a by-product of further incitements to consumption. If citizens want splendid cities and clean streets, then bond issues can pay for gardeners and attendants and cleaners, all better paid and in prideful uniform, officials of an urban culture they are glad to serve.

These are simply a few instances of the way in which nonpollutive service activities can provide valid jobs and valid growth without putting a single particle in the air or an effluent in any river.

In short, growth and environment are not in necessary opposition.

If population becomes stabilized, basic injustices are redressed, effluent charges imposed, new technologies of nonpolluting technologies evolved, the pressure of arms relaxed, and citizens persuaded, by education and example, to widen the range of their nonconsumptive joys, societies can still "grow," yet still preserve and enhance their environments. If these seem a lot of answers to a single issue, this is the way, in nature, that answers are given, by a myriad of approximations and interdependences which keep the whole system, short of catastrophe, in some kind of dynamic balance.

At which point the impatient citizen may well exclaim that he is not interested in Utopia. He does not want the new Jerusalem. He simply wants life as usual with a little less dirt and a little better garbage disposal. But the difficulty is that this straightforward pattern of sharply rising and disguisedly subsidized consumption which has been for fifty years at least the goal of most developed societies—and the only tool some of them have used to deal with social conflicts—cannot be extrapolated without change into the twenty-first century. It has already produced self-confessed "monstrosity" in Japan, and in many other lands less courageously ready to give public recognition to the fact. With the whole economy, as with its most vital strand, energy, there will not be enough air, water, space, and amenity left for two and a half billion developed peoples still increasing by at least 0.5 per cent a year unless they adopt a more modest attitude to purely material demands, unless they invent forms of consumption and enjoyment that make fewer claims on a limited biosphere which is all there is to support them and which, by the year 2000, will be supporting five billion other people as well.

Here, then, can be the lasting goals of a more sustained and successful environmental movement. Today as we look back over the longest, most consumption-oriented boom in man's brief industrial history, we must ask whether, as with the earlier efforts, the movement of protest and interest is destined to peter out or whether, on the contrary, we are genuinely reaching a new level of sustained awareness in our ecological thinking.

Within the developed world, there are, surely, reasons for hope. In the first place, the arguments for more restraint and care are very much more pressing than they were in earlier periods. In developed lands so many basic human needs and satisfactions have been catered for at a rising cost in environmental damage that in the strictest sense benefits "at the margin" are beginning to change. In order to get the majority of the people out of abject misery, it may be worthwhile tolerating a good deal of smoke and dirt. To add another car and another television set to already affluent families at the cost of dead lakes and dying rivers begins to look absurd.

In the second place, there does appear to be somewhat less division between the various citizen groups interested in a better environment. It is true that particular people remain wedded to particular priorities—to family planning, to model cities, to the wilderness, to threatened whales. But the gross misunderstandings, the total lack of common interest between middle-class groups caring about trees and bald eagles and urban reformers bent on getting lead poisoning out of slum children is surely less divisive and counterproductive today than it was even fifty years ago. The chances are that a citizen who wants a cleaner river may also want low-cost housing and urban parks. A reformer who begins with genuine urban renewal will not find it strange that control of land use extends to recreation areas and out to the wilderness. Divided interests may still make the environmental movement less potent than it might be. But for the first time, the logic of convergence is beginning to outweigh the effects of sects and schisms.

Perhaps the reason for this is, in the last analysis, the emergence of a science of the environment. The balance of man's scientific drive and interest is beginning perceptibly to switch, especially among the young, from the vast and heady triumphs of physics and engineering—speedier machines, more powerful rockets, deadlier weapons—to the deciphering, with all the exact patience that is required, of the minute balances, the tiny trade-offs, the intimate structures of living things in their complex, vulnerable, and interdependent ecosystems. It is the shift, if you like, from the immense power that can be mobilized behind the valid generalization to the exact knowledge that can flow from the particular instance, the shift from big to small, from quick to slow, from the forceful to the delicate. Such an alteration of balance may not match the appetites and energies of modern man. But compared with the intellectual basis of earlier environmental movements, it has a depth, a cogency, and a consistency, all of which could be precursors of a major shift in man's national interest and scientific approach.

These are the reasons for hope. But the picture is not all of hope. With population piling up, resources growing more expensive, technologies more complicated, and still, in all societies, human aspirations on the rise, we have little time in which to make our more responsible choices and better environmental judgments. Above all, whatever their good will, most developed peoples are still affected with one type of "tunnel vision." Although they make up no more than a third of the human race, they find it exceptionally difficult to focus their minds on the two-thirds of humanity with whom they share the biosphere. Like the elephants round the water hole, they not only do not notice the other thirsty animals. It hardly crosses their minds that they may be trampling the place to ruin.

It is here in the extra dimension of concern for *all* the planet's peoples that the environmental movement still displays the lack of coherence that fatally and periodically weakened it in the past. Yet today it is precisely in the developing world that some of the most contradictory and destructive problems of the human environment are to be found.

Part Four: The Developing Regions

10 PROFILE OF DEVELOPMENT

An Uneasy Inheritance

WHEN WE SPEAK OF developing countries, the phrase has nothing to do with levels of culture or history or contribution to mankind's heritage of civilization. The phrase in the main means simply that a society has not yet crossed the threshold to the modern, high-technology society with all the advantages and evils this passage entails. The category includes countries of immensely old and sophisticated civilization, such as India or China—which, between them, make up a third of the human race—long established literate and urban-oriented societies in Latin America and some of the most ancient and continuous of all the world's political units—Egypt, for instance, or Iran.

A fairly arbitrary estimate is often made which fixes at an annual income of $500 a head, the level at which a country begins to emerge fully from the pretechnological condition. But 80 per cent of the nations at and below that level have annual per capita incomes of less than $250. This figure gives a better guide to the bleak realities of personal poverty for citizens and straitened resources for governments in developing lands. Investment to provide for the essential increases in productivity needed for development is a third lower than in high-income countries with incomes above $1000 a year. A third of this investment is not covered by local savings and must be secured abroad. Tax revenues, another critical source of funds for investment, welfare, and amenities are only just over half the level in developed lands.

All this is a statistical abstract of extreme shortage of resources, either private or public, for consumption and for investment. The profile gives us other evidence of pressure. For instance, nearly 60 per cent of the population is still working in agriculture, where productivity has been, until very recently, almost universally low. (The figure for the most highly developed nations—above $2000 per capita—is only 8 per cent of the work force.) Nearly 70 per cent of exports are still primary products, which (with a few exceptions like petroleum) tend to fluctuate most widely and weaken most easily. Yet imports of goods and services, mainly the expensive machines and skills needed to increase productivity, are not much lower than those of the developed nations already at the $1000-a-year level. Urban population jumps nearly eight times over in the passage from $100 to $200 annual per capita income and is already double the number of people engaged in industry. In the $1000-a-year group, there is only a 15 per cent difference. School enrollment and literacy have shot up to cover over half the population and whatever the dropout rate, this is a formidable force building up for employment outside traditional, illiterate, and subsistent ways of life.

Finally, growth of population, at an average of 2.5 per cent a year, is a third to a half higher than in developed lands or during their nineteenth-century development. The figure gives a 2 per cent annual increase in the work force, double the nineteenth-century rate. And these are averages. In some areas—the Maghreb, Iran, Central America, Andean Latin America (except Chile)—the growth in population has risen 3 per cent and more.

It is not difficult, therefore, to understand the driving dedication of governments in developing countries to get their peoples out of a trap of poverty more locked and complicated than any experienced in earlier times. Each obstruction tends to reinforce the next. Population and work force explode ahead of industrialization. Migrants pour into the cities. Industry, often under foreign direction and ownership, introduces modern labor-saving technologies when unskilled labor is chiefly available. Markets overseas are blocked by the presence there of monsters of efficiency—Mitsuis, IBMs, Volkswagens—and by the tariffs raised to keep out cheap labor-intensive goods. Markets at home remain limited by local poverty or, often enough, by the extreme smallness of the post-colonial states. Such are the difficulties that make up a maze, a web, an obstacle race for developing governments which both intensifies their determination to break out of poverty and frustrates the efforts which they have to make.

In developing as in developed lands, all the pursuits and consequences of the only exits from poverty—greater productivity, the "more for less"—have their impact on the environment. More productive agriculture brings the familiar problems of fertilizer run-off and pesticides. Industry puts its effluents into the free goods of air and water whether the industrial area is the Damodar Valley in India or Tokyo Bay. Urbanization concentrates all man's wastes in unmanageable amounts, whether we talk of Paris or Bangkok. But we have to remember that in a number of significant ways, the inadequacy of purely economic calculations, the environmental risks, the calculus between profit and loss in the broadest sense of economic and social accounting are at once different and more difficult for those who are the latecomers to the technological revolution.

In the first place, the urgency behind the need for economic growth is universally greater because of population pressure and intense poverty.

Secondly, ecological risks are different and possibly more damaging in a rapid expansion of output in farming because soils and climates are less benign in tropical regions and because far less is known about their structure and possible response.

Thirdly, the external diseconomies are subtly different. In many areas, in particular where industrialization lies mainly in the future, the purely physical pollutions—of overloaded airs and waters—are not yet the most serious. But universally the external diseconomy which could come from driving blindly for economic growth without considering the social implications of employment, income distribution, internal migrations, and exploding cities are already immediate and potentially catastrophic. In the urban areas, in particular, anarchy and breakdown may well be called the chief ecological risk.

Finally, at that fluctuating but vital point where economic growth and national self-interest either reinforce or defeat each other, developing nations face a unique challenge. On the one hand, economic development to satisfy their peoples' aspirations, particularly in the immediate post-colonial situation, is the very stuff of effective political leadership and national identity. On the other, so closely are all developing nations plugged into the international circuits of a world economy which they did not create and of which they are still relatively powerless members —commanding only 25 per cent of its resources for 75 per cent of the planet's population—that greater efforts at growth, particularly at growth with sound environmental safeguards, places them before a dilemma. Do we, they must ask, seek further involvement, further dependence so as to secure the desperately needed resources fast? Or do we cut ourselves off—like Japan in the seventeenth century, America in the eighteenth, Russia and China in the twentieth—and go it alone, risking thereby a worsening of our peoples' precarious condition and in fact postponing that ultimate balance between independence and interdependence which can be the only finally sane objective in our inescapably small and totally involved planetary existence?

As we look at each of these issues—population pressure, the modernization of agriculture, industrialization, the plight of the cities, and the relations between the developing south of our planet and the developed and wealthy north, we shall encounter again and again the fact that action is possible, policies are available, success is conceivable but only on two conditions—that all external diseconomies are taken into account, including the social ones, and that the effort to achieve a more effective and more balanced development is not defeated by the inability of all the world's nations, rich and poor together, to mobilize sufficient capital for the task.

Population Pressures

The first, most obvious, and widely publicized pressure is that of population. In this century, a very great increase in the control of major epidemics—yellow fever, smallpox, plague, and, particularly since the Second World War, malaria—has brought death rates sharply down all round the planet in societies which were not being transformed at comparable speeds in any other area. The classic instance is Ceylon, where the death rate was halved and population doubled all in the space of twenty years.

The general rate of population increase—of over 2.5 per cent a year— is not only unprecedented in human history. It promises further increases of an all but inconceivable kind. Each expanding generation leaves an even larger base for the next expansion. The present 2 billion peoples in the developing world cannot fail to reach 5.5 billion by the year 2000. If there were no change in policies and life-styles, the 2020 figure could be 14 billion. By 2050, it could be 28 billion. No one can give a precise estimate for the ultimate numbers of people our planet can support. It depends, of course, upon the standards at which they want

POPULATION ESTIMATES:
INDUSTRIALLY DEVELOPED AND DEVELOPING COUNTRIES

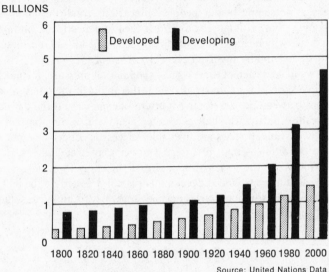

Source: United Nations Data

to eat, be housed, educated, made healthy, and move about. In any case no one denies there is a limit and if in the less industrialized world the mathematics of growth at the present 2.5 per cent (occasionally 3.5 per cent) growth rate continues unchanged, we will be adding to our planet a billion people a year by the end of the next century, at a date not much more distant from us in the future than we are from the opening of the first Great International Exhibition in London in 1851, which was designed to demonstrate the unlimited hopes for peace and prosperity inherent in the new industrial system.

The point at which, in the next century, the rate slows down and begins to stabilize is thus a cardinal problem for all the world's peoples. Demographers have estimated that if the developed peoples reach an average family size of two children by the year 2020 and the developing nations by 2040, total world population would level off at just under 16 billion. If twenty years could be taken off this adjustment, total population might stabilize at just over 11 billion—or about three times its present level. This is widely held to be the limit for a reasonably well-fed and healthy but by no means luxurious life for all the world's peoples. At some indeterminate yet unavoidable point beyond that scale, the old forms of birth control—disease, famine, war, and death—would restore, in traditionally brutal form, the balance between fertility and mortality.

In theory, there is no inherent reason why a stable world population should not be reached by less inhuman means. Modernized societies have drastically decreased family size as a result of basic changes in their profile of development. The first is, paradoxically, the survival of children. No longer is the annual pregnancy needed to make absolutely certain that even three or four children live to be adults. The chief factor in the last century's halving of death rates is simply that more babies live. Alternative employment outside agriculture has lessened the need for strong sons to till the fields and obedient daughters to care for the animals, the vegetables, and the weeding. In an urban apartment, twelve children make no direct contribution to family resources. Another factor is a lessening of dependence upon family support. Alternative schemes —insurance, welfare—have arisen to help in providing for old age. The education of women opens to them other horizons than wholly domestic occupations. Family life on an extended scale is no longer the chief source of entertainment and solace. All these factors, without any public support for family planning or for abortion under proper medical supervision, have lessened family size throughout the developed world, quite irrespective of culture, ethnic background, or religious faith. The birth rate in Italy is lower than in Romania, Spain's is at the same level as that of the Soviet Union.

What we cannot tell today is whether the addition of a new factor —intensive government education and action in favor of family planning —will speed up the decline of family size while all the other changes in modernization still lag behind. A family migrating to the United States at the turn of the century could come from an area with a fairly high birth rate—rural Sicily, for instance—yet conform within a generation to the pattern of smaller family formation already beginning to dominate America's urban culture. But in the developing world today, there will still be over 2.5 billion people living in rural areas in 1980. In the year 2000, the figure is likely to be 3.5 billion. Moreover, the 3 billion people who by 1980 will be living in urban settlements will include so large a percentage of first-generation rural migrants that city attitudes and habits in the "developed" sense may not be quick to influence family size.

It follows that two factors of equal importance are involved in the critical issue of slowing down the world's present untenable speed of population growth. The first is a new factor—strong governmental policy in favor of smaller families. The second is the older, more complex but already fairly successful solution—the change of the whole context of people's lives into the more modern habits of high education, female emancipation, rapid industrialization, productive, modernized agriculture, and city life. On ecological principles, we may guess that the second solution is likely to be more effective, since fully sustainable changes tend to result from interacting, complementary, and mutually reinforcing developments. It is surely significant that those developing countries in which the birth rate has fallen most sharply have combined governmental policies in favor of family planning with very rapid rises in agricultural modernization, industrial growth, and urban expansion. Government policy can be a vital part of this process. But as a single and specialized type of intervention, it needs the development of the wider context of modernization to accomplish its full aims.

Both approaches present difficulties. The first is the differing degrees of immediate population pressure experienced in various developing countries. In tropical Africa there are nine countries with a density of less than 10 people per square mile, twenty-seven with less than 50—the figure for Europe is roughly 150. Similarly in South America there are no countries with more than 50 people per square mile. Some of their governments feel that only rapidly rising population will enable them to fill up their national boundaries, achieve the "critical mass" of producers and consumers needed for a modern economy, or produce the number of citizens needed for prestige, power, and successful national statehood.

In most of Asia, where the world's great concentrations of population have been growing for over two millennia within the framework of high civilization and stable traditional agriculture, these considerations do not apply and it is significant that the two largest states, India and China, both have active governmental policies designed to stabilize their population. Indeed, China appears to have reduced its rate of increase—until recently producing 15 million more Chinese annually—to under 2 per cent a year.

These examples are encouraging, for they represent the only conditions under which strong governmental policies will be introduced. This is the perception, *by developing governments themselves,* that the pursuit of high population is as ultimately disastrous for the nation's well-being as a failure to try to increase productivity in farming or to introduce modern industry. No amount of rational or well-meant advice and offers of assistance from other governments or agencies can be effective until this essential internal decision has been taken. It can be particularly unwelcome when it is given by countries who, with less than a third of the world's population, consume over 75 per cent of the world's income. But China and India are seeking to discourage large families in terms of their own self-interest. Some twenty-five governments in developing countries are adopting the same internally and rationally determined approach. The reasons were very clearly stated by one African government in 1970—in a country which is still free from any direct population pressures of an Asian kind. On launching this country's Family Planning Program, the Finance Minister of Ghana remarked:

The present rate of growth increases our population by 5,000 people every week. . . . In simple terms, it means that as a Nation we are increasing in number faster than we can build schools to educate our youth, faster than we can construct hospitals to cater for the health needs of the people, and faster than we can develop our economy to provide jobs for the more than 140,000 new workers who enter our labor force each year. Our rate of population growth thus poses a serious threat to our ability both as individuals and as a government to provide the reasonable needs of our people. . . . Thus we see that our population growth and our reproductive habits pose very serious problems which must be tackled realistically and effectively NOW if we are to avoid the justifiable curse of our children and those who come after them.

We are aware that there will be some in our midst to whom these dangers are more imaginary than real. There are those in the grip of the dangerous illusion that the vast expanses of underdeveloped land invalidate the argument for the regulation of population growth in Ghana. They fail to realize that invariably the land remains undeveloped because of the lack of capital and technical skills required for its development. There are also those who still cling to the equally dangerous misconception of the prestige value of large populations in a techno-logical age when the quality of our people is more important than their numbers.

This is the essence of the problem. In the nineteenth century, the growth of population was less than 2 per cent, the labor force did not increase by more than 1 per cent. At these rates and under prevailing conditions of much simpler technology, the growth was a positive stimulus to modernization. More productive agriculture needed fewer workers to provide the same amount of food. The workers thus released were transferred into industrial work with still higher productivity. As a result, the amount of work achieved, the amount of potential consumption that followed, stimulated the overall growth of wealth.

The process, though painful and disruptive, did give a measure of economic momentum. But at 3 per cent rates of overall growth and a 2 per cent increase in the labor force, the balance between the amount of capital needed on the one hand to educate, train, and house the workers and, on the other, to invest in more productive farming and in industrialization simply gets out of line—as it is in large parts of Latin America today. The untrained worker is not a net addition to a productive labor force or to a lively consumer market. He produces so little that even his minimal consumption represents an economic loss. He makes no contribution to his country's growth or strength. On the contrary, he becomes yet another pitiful "marginal man" stranded on the edge of a less than productive agricultural system or joining the ranks of the unemployed in the squatter fringes of big cities.

In fact, there are strong economic grounds for arguing that if capital can be effectively invested in the policies, services, and techniques which help to slow down population growth, the result can give a larger and speedier net benefit than comparable investments in, say, a steel mill or a petrochemical complex. Capital is, after all, savings. In other words, it is nonconsumption. It can be achieved either by a very rapid rise in productivity—the production of more goods for the same inputs, the balance being saved—or by a reduction in consumption. But a very rapid rise in productivity does not accompany the earliest and most basic investments in an economy. Roads, rails, dams for irrigation, power plants demand the input of big lumps of capital and do not immediately release a large increase in resources. But if consumption can be held down by a smaller increase in the number of mouths to be fed, the claims on resources are automatically and immediately less. Thus public spending on a program which decreased family size by a given percentage could give a higher return than the same amount invested in capital works or in all or most resources regarded as essential to the country's capital infrastructure.

The perception of such potentially beneficial results is a matter for the self-education of governments. There are at least signs that within a decade, a population policy may be as much a symbol of enlightened modernity as the expansion of power and transport or the introduction of better seeds and fertilizers. One may even stretch the imagination and believe that it could be seen as a more rational approach to the citizen's well-being than the purchase, say, of supersonic fighter-bombers out of dwindling supplies of foreign exchange. Yet it is clear that any change in government policy will be a hundredfold more successful if at the same time there is a radical transformation of the whole traditional social context in which high birth rates were a response to the high mortality now banished by modern control of disease. In other words, the most successful form of population policy is effective development. In modernizing agriculture, in speeding up industrialization, in building the urban infrastructure, in creating new jobs and opportunities, governments are also encouraging new family attitudes and stabilizing family size. The only problem is the cost and scale of the whole program.

11 POLICIES FOR GROWTH

The Green Revolution

WE BEGIN WITH AGRICULTURE because, as governments in most developing countries now realize, it is the basis of everything else. Until quite recently, the chances of food supplies in developing lands, particularly in the densely populated areas of Asia or the arid lands of the Middle East, looked disastrously unpromising. The starting point was dismal enough. The bulk of the developing peoples with per capita incomes below $200 were eating less than 2000 calories a day and the critical intake of protein was, on the average, not much above half that of the admittedly occasionally overstuffed inhabitants of developed countries. These levels of consumption are simply insufficient for full health. The Food and Agriculture Organization has estimated that 300 million children in developing lands have "grossly retarded physical growth." Nor can we forget the retarding effect of protein deficiency on mental development.

But the existing low base of nutrition is only the starting point. Between 1960 and 1965 the alarming fact emerged, in the wake of worldwide population censuses, that while population in the developing countries had grown by 11.5 per cent, their food supplies had increased by only 6.9 per cent. The gap was actually widening and with unchanged agricultural practices would continue to do so. In 1962 an average farming family of between five and six people had about 6.5 acres on which to feed themselves and just over two and a half other people. By 1985 they would be down to 5 acres and the others dependent on them would be just over four. This is the inexorable mathematical progression of population growth. Yet it is virtually impossible for traditional agriculture to go on supporting such an increasing load of people on a limited amount of land.

In these societies—whether they are the tribal lands of Africa or the vast feudal estates of Latin America or the small scale landlord-tenant structures of Asia—the fertility of the soil is maintained principally by allowing it to rest (under fallow) and to regenerate its own nutrients. As population pressure increases, the amount of land that can be left lying idle must shrink. So do the woodlands which have not yet been cleared for cultivation. But as fallowing times are reduced and forests are felled, the natural recovery of fertility is cut as well unless nutrients are added by the cultivator. Output declines, land is exhausted, it begins to erode and wash away or bake to bricklike hardness. At the worst, man-made deserts begin to form—as in Rajasthan or North Africa—and the drift of their dust and the changes they bring in overall humidity reduce further the amount of available farmland, increase the pressure of population, and thus extend the damage. Overgrazing of animals has the same effect. Indeed, many nomad tribes should be called not the sons but the fathers of the desert. Provided no sharp changes are made in arable or pastoral habits, developing nations, particularly in Asia, could by the year 2000 be heading back to the old famine cycle.

However, in 1967 there came a breakthrough in agricultural productivity, the so-called Green Revolution. Twenty years of research went into producing new, carefully selected hybrid strains of rice and wheat which could safely absorb up to 120 pounds of nitrogen per acre. Traditional strains could do so but the resulting heads of grain were simply too heavy for the thin stems and simply fell over or lodged if more than 40 pounds of fertilizer was applied to the acre. This increased tolerance for fertilizer, combined with a quicker period for maturing—only 120 days compared with 150 to 180 with older species—makes the new hybrids two or three times more productive, provided they receive enough water, fertilizers, and pesticides. One or two examples of increased percentages of growth can illustrate the scale of the change. Between 1954 and 1964, for example, the production of Asian rice, the continent's staple food, grew on average by only 1.4 per cent a year. The new strains can give a growth rate of 2.7 per cent, slightly ahead of population expansion. The figures for hybrid wheat are even more remarkable. Asian wheat can increase from 0.8 per cent to 4.3 per cent; Near Eastern wheat from a minus quantity—a shrinkage of 0.8 per cent to a plus of

2.8 per cent. If these increases are secured, on a sufficient acreage planted to the new strains, some of the areas of highest growth in population have some prospect of feeding their growing populations well into the eighties.

For Asia, this breakthrough has a special significance. Much of the growth before 1965, especially in the Indian subcontinent, came from bringing more land into cultivation—land which was often of marginal quality. But it is estimated that India is already cropping 402 million acres out of a potential 410 million. Mere extension of cultivated land has ceased to be an answer. The relatively large increase in new acreages under crops that the Food and Agricultural Organization (FAO) has proposed as targets for the seventies and eighties—from 1400 million acres to 1650 million acres—will have to take place very largely in Latin America and Africa. And even with this reserve brought into use, the average man-acre ratio of developing countries is bound, as we have seen, to fall steadily over the next three decades. The only answer to this degree of pressure is a radical revolution in farming methods, which by increasing productivity feeds the farmers, provides food for the workers moving to the cities, underpins industrial expansion, and helps to provide a lively internal market for manufactured goods.

The agricultural breakthrough of the Green Revolution can hopefully set or keep in motion these essential elements in the overall momentum of modernization. But in doing so it will confront developing peoples with some special environmental problems. In terms of immediate action, the first point to recognize is that most of the world's careful agricultural research has not been concerned with the special conditions of food-growing in developing lands. Virtually until the last decade, it concentrated on the problems raised by the foodstuffs grown in developed countries or on those of the agricultural raw materials—coffee, tea, cocoa, sisal, jute—exported to them. A very large part of the research needed now is for food supplies grown in nontraditional ways on the soils and in the climates of the world's tropical and subtropical belt. These areas differ radically from temperate regions in three ways. In many of them, soils are very fragile. The humus is provided largely by leaves falling from the dense, varied, balanced biomes of the rain forests. Any injudicious clearing of the trees exposes the undernourished soil to Niagara-like downfalls of tropical rain and leaches out the remaining nutrients. When the rains end, either the winds come and blow away the topsoil or the equatorial sun bakes the soil into bricklike laterite. It is estimated that 92.8 tons of soil were lost for every 2.5 acres in a region of the Ivory Coast that was cleared for growing cassava, while only 1 ton was lost per acre in an adjacent secondary forest in 1956. Similarly, an area in Senegal cleared for peanut cultivation experienced an erosion loss of 14.9 tons per 2.5 acres, whereas the loss was only 0.02 tons per 2.5 acres in a nearby forest.

Flooding rain is the second difference. The gentle, steady—and occasionally irritating—year-round rains of temperate climates are nature's least chancy and disruptive gift to man. Farmers are not so well served in the tropical belt. When the rains come, they come with the force of an innundation. Yet for six months of the year, they may hardly flow at all. The monsoon rains in India blow in from the Indian Ocean and the Bay of Bengal fairly reliably at the beginning of the summer and south India can normally be sure of replenishment. But as the rains travel inland, they shed their weight and lose their momentum. In the Indus Valley system there is some abnormality—a degree of drought, some local flooding—in one year out of three. Then, in some years, the rains fail and there is a complete drought.

To crown these fragilities, temperatures of over 100°F can be recorded in the hottest regions and since they are usually arid and thus most in need of man-made schemes for providing water, evaporation is a constant risk with its dangerous consequence of moisture vanishing, leaving only salts behind. From these facts—the delicacy of the soil, the nature of the rainfall, the scale of temperatures—there follows, inevitably, a wider range of environmental risks than in the climates of temperate lands. It is this fact of larger risk that requires a wholly new thrust of agricultural research, testing, and feedback.

The first priority—both for research and practical policy—is the problem of supplying regular water. It is quite useless to invest in the whole apparatus of the Green Revolution and have the expensive inputs —improved seed, fertilizers, pesticides, trained manpower, enlarged facilities for storage and transport—nullified by the failure of the rain to arrive on time.

We begin, therefore, with irrigation. According to U.N. estimates in the mid-sixties, the areas where population pressure and uncertain water supplies are most critical—India and China—have already gone furthest in developing their potential for irrigation. Although at least half Asia's irrigation comes from minor schemes and there has been a phenomenal expansion of tube wells in the Indian subcontinent in the late sixties, the largest environmental impact is obviously exercized by the bigger schemes which involve building dams and barrages in major river valleys together with main canals, feeder channels, and irrigation ditches down to the level of the farm.

In these big schemes, two difficulties stand out. Earlier schemes, planned solely with the aim of irrigating farms and increasing food output, were developed without a full understanding of the complexity of possible interactions between soil, water, vegetation, and climate. For instance, the vast irrigation works in the Punjab, which harnessed the waters of the Indus for use in human settlement and irrigated farming at the turn of the century, aimed above all at drought insurance. They were designed to ensure some water to every farmer, even in a dry year. The canals, the channels, the feeder ditches spread water as far as possible and it took half a century to learn that this type of distribution—giving a kind of minimum flow to a maximum number of users—would little by little undermine the effectiveness of the whole system.

From the far-flung channels there was steady seepage, raising the water table constantly until it began to damage the plants' roots and create a risk of waterlogging. But the very light surface flow was too modest to flush away the salts which settled on the soil after intense evaporation in late spring sunshine. The fact that the whole Indus basin had once been an inland sea only increased the risk of salinization. By the end of the 1950s, Pakistan government officials in Rawalpindi would point out that land the size of a tennis court was vanishing from cultivation every twelve minutes into the white, crusted sterility of salinized soil. It was, if you like, an archetypal example of that kind of human intervention which, beginning with good will and apparent technical efficiency, ends in potential disaster because the complexity and delicacy of nature's ecological balances have not been understood.

However, in this case, as in many others, further knowledge could be drawn on to correct the original miscalculations. In the early 1960s, an expert team visited the Indus basin, analyzed the underlying causes of increasing sterility and salinization, and recommended a solution which was as effective as it was straightforward.

In most of the affected areas, the water table contained fresh water since it was being constantly reinforced by the rivers pouring down from the Himalayan snows and from the seepage of river water through the lining of the canals. If this underground water could be tapped by tube wells and poured across the salinized soil in sufficient quantities, the water table would be lowered, the salt washed off, the earth cleansed, and harvests restored. At the same time, higher dams upstream on the Indus, including the large Mangla Dam, would hopefully ensure a future flow strong enough for an adequate flushing of canals sufficiently well-lined to prevent constant seepage.

Within half a decade, land which had fallen into a flaky scrofulous whiteness, a sort of leprosy of the soil, began to recover its fertility. The brilliant green of growing wheat reappeared across the fields. Farmers hastened to install tube wells. Sixty thousand appeared in a couple of years. True, there are still problems. Desert areas in India and Pakistan are still spreading. Not all the ground water is fresh and some areas of salinity have not been cleansed. Yet the experience has shown that the mistakes made at one stage by ignorance can still be repaired, before final disaster, by the application of wider knowledge and a keener sense of the interactions of natural systems.

A second and quite different kind of mistake has been made on a number of sites in the great surge of dam-building that began in the developing world after the Second World War. The prime object was not water for agriculture. These were the decades of maximum emphasis upon the necessity of rapid industrialization and hence on the overriding need for electricity. The big dams of the 1950s and 1960s—the Damodar scheme or Hirakud in India, the Volta and Aswan dams in Africa, the Mekong proposals for Southeast Asia—were all primarily conceived of as colossal engineering feats which could make available, on the lines of such prototypes as the Dneproge or Tennessee Valley systems, vast amounts of electrical power produced in what is, without question, the least immediately pollutive fashion.

Given the violence of the climate in many of the chosen areas, defense against flooding was thought of as an extra bonus. So was the chance of easier transport by water along lakes and more stable rivers. Often as afterthoughts, the planners totted up some gains from tourism. But apart from the great Aswan Dam, no very large component for irrigation was included. Most large dams were built to deliver massive quantities of electricity to industry. They were designed by engineers and although some of the preliminary surveys took the greatest possible care to consider such problems as the resettling of the people who would be flooded out in the expanding reservoirs, the effect on downstream fishing, or the creation of new microclimates in the region, ecological studies of the effects of the new water bodies in their totally interdependent natural surroundings were not wholly systematic. This was hardly surprising since engineers and planners had not been trained to grasp the full meaning of natural complexities and interactions.

The major risks which have begun to emerge as a result of some of this dam-building affect not simply agriculture but, ultimately, the power supply itself. In a number of cases—in the Damodar Valley, for instance —the effect of flooding agricultural land is to drive subsistence farmers further up the slopes of the valleys. There, in order to survive, they are compelled to carry on the traditional practices of shifting agriculture, slashing and burning the trees, cultivating the cleared patch for a year or two, and then moving on leaving the forest in need of fifteen or twenty years of natural regeneration. But on steep slopes, the clearing of the land hastens soil erosion as the monsoon lashes down. Moreover, the resettled people may be doubling up with existing farmers, cutting the fallow time, and recultivating the land before it has replenished its nutrients. Such soils are increasingly brittle and begin sliding off the rock on which, over billennia, the patient work of crumbling roots and breathing plants had built them up. Soil, falling away from steeper slopes, slips into the reservoirs. Silt which can no longer be washed downstream begins to fill the bottom. Silt is in fact the major pollutant released by farming. Unless the degradation of the watersheds can be checked, not only will the hills turn to stone and laterite. The flow of water and hence of power from the dam will peter out.

Sometimes the troubles are downstream. Below the Kariba Dam, the Tonga tribe cultivated intensively the land between the Zambezi's highest and lowest flows. The alternation was seasonal and reasonably reliable. The whole cycle of local work was geared to these rhythms. Once the dam was built, the release of water came to depend upon the needs of electricity generation, and releases occurred with no particular rhythm, no flow alternating with vast flows, bumper crops with no crops at all. The loss of land for food in this case does not appear to be great and Zambia, at least, is not short of land. But in more crowded areas, electric power might finally be purchased at the cost of a dangerous fall in cultivable land and reliable food supplies.

A comparable problem arises when the undammed river has washed down rich silts, fertilizing the flood plains, building up usable land at the river mouth, and even adding nutrients to estuarine waters. The installation of dams and barrages may lessen this flow with multiple consequences—more silt building up behind the dam, less fertile land shrinking deltas, and fewer fish in the neighboring sea. Some diminution in the sardine catch in the eastern Mediterranean and the risk of diminishing fertility in the delta of the Nile are cited as less desirable consequences of the construction of the Aswan Dam.

Another whole range of problems concerns people's health. The tsetse fly, which afflicts cattle and even humans with trypanosomiasis (a sleeping sickness), appears to be particularly well served by habitats where a great many different kinds of vegetation are mixed together. Such a prolific mingling is encouraged by the rising waters of reservoirs. Another lies in the spread of bilharziasis (or schistosomiasis) in the wake of more permanent water supplies. The trematode worm which causes acute discomfort and lassitude in adults and can kill children spends part of its life cycle in a waterborne snail. If channels are dry for part of the year, the snails have a hard time of it. But the kind of perennial irrigation which allows double and triple cropping of grains also allows the same multiplication of snails. If, at the same time, sanitary habits are primitive and human excrements simply washed away into the ditches, the worm completes its life cycle from snail to intestine and back to snail. The problem is already a bitter one in Egypt. In China, schistosomiasis ranks among the ten most important diseases. It is estimated that, world-wide, more than 200 million people are affected.

Are we then to conclude that the critical input of steady water supplies needed for the "miracle grains" is being purchased at too high a price? Must declining health in humans and increasing sterility in the soil follow the construction of the larger schemes? Are we once more on the treadmill—more people needing higher food supplies, more food, demanding more water, more water, more dams, more dams, more damage?

This whole progression is based on too simple a calculus. The dams do, after all, provide the power they were planned to produce and no one denies that the generation of large supplies of energy is needed to develop industrial and urban sectors into which the growing rural populations can be drawn and thus relieve the pressure on the land. It is also essential to any *regional* development of modernized farming or of decentralized economic growth. To talk about the risks of bilharziasis while failing to mention the benefits of kilowatts of power does not give the true balance of advantage.

Besides, there need not be a fatal connection between valley management and increased environmental risks. Many of them are due to the fact that the large schemes are too often planned without proper weight being given to complex ecological factors. The measurements and calculations of the engineers—in terms of stress, of needed materials, of on-site management, of maximum and minimum power potential—are all precise, exact, and excellent. But, thereafter, too much has tended to be guesswork. In fact, there were cases in the Indian subcontinent in the 1950s where even agriculturalists were not drawn into the planning of the big dams. The installations would be completed and main channels built by the well-trained and confident corps of hydraulic engineers before local ditches had been put in, farm-to-market roads constructed, storage built, soils tested, or crop patterns and varieties discussed. Water would be standing by while peasants continued dry farming.

The lack of sanitary and ecological information follows the same pattern. There are simply not enough trained people and not enough organized and tabulated material to ensure that the kind of miscalculations about natural balances and interactions suffered on one dam site are available and studied in the preparation of another. Take an instance from the field of sanitation. The effects of bilharziasis are not unknown. The kinds of sanitary regulation needed to lessen the risks of infection are at least as important, in terms of human welfare, as the stresses of concrete in the dam or the safe transmission of high voltages of electricity. It is just as inconvenient to be killed by a sanitary failure as to be drowned by an engineering one. The only clear advantage is that the drowning may be quicker. Yet not one tithe of the expertise, care, planning, expenditure, and feedback go into the health aspects of irrigation as into the construction of the dam.

Or take another less catastrophic but nonetheless important issue. The hope of large fish catches in the new reservoirs is always thrown in as a useful, protein-giving extra when big dams are discussed. But the lake behind the Kariba Dam has shown a pattern of a sudden increase in fish population and then a drastic fall. No one is quite sure at which point a balance between algae, edible fish, and various predators will be established. Moreover, both here and on other schemes, little thought seems to have been given to the types of fish that should be introduced, to the training subsistence farmers need to shift to fishing, to the storage and transport of their catch, and to the availability of local markets. Yet without these studies, as detailed and careful as the measurements of the dam itself, the total ecological cost of large-power schemes, now and in the future, will nullify some of the undoubted benefits they were designed to produce.

These environmental blind spots can be illustrated from another aspect of the Green Revolution. We will assume enough water for a big advance in the use of the new hybrids. But the new types of wheat and rice have been carefully bred above all to produce maximum yields with heavy use of fertilizers, together with strong resistance to the most widespread diseases. They have *not* been closely related to local climates, lengths of sunlight, growing periods, pests, soils, and all the myriad conditions which make up the plant's natural local habitat. They tend to be monocultures with the inherent fragility that comes from lacking a variety of possible responses to unsuspected challenges from the environment around them. The seeds they replace have lower yields but very probably greater tolerance since they evolved over millennia their ability to defend themselves against local pests and diseases.

Do we then conclude that the spread of these new seeds, so spectacularly successful in many places, is preparing an ultimate catastrophe in which the ricelands and wheatfields of Asia and the new corn sown in Latin America will be swept toward irresistible damage, leaving famine in its wake? Clearly, the answer cannot be dogmatic. The monocultures of grain in North America have fed the local people and much of the world for over a century with considerable success. True, the soils are richer, the climate kinder. It is also true that in western Canada, different varieties of high-yielding grain have been alternated to keep rusts and predators guessing. But it does not follow that the answer is simply to slosh on the latest invention in pesticides or herbicides. One of the classic and most quoted examples of this comes from Peru where DDT and other chlorinated hydrocarbons were used to spray the cotton crop in the Cañete Valley. The sprays killed the pests. They also killed the killers of the pests. By the time the original pests had developed DDT resistance, all their old beneficent enemies had vanished. The pests multiplied, even more applications had even less effect, and before long the whole crop was in jeopardy.

However, the Cañete Valley case history has a happy and significant ending. Overall spraying with all-purpose insecticides was abandoned. Beneficent predators were introduced from outside, planting and harvesting were timed to avoid the times of infestation, highly specific insecticides were added to the total "system," and as a result, the cotton crop recovered. A single-shot effort was abandoned in favor of a mix of natural, artificial, and also traditional methods. The more complex answer provided a solution whereas the cruder interventions did not.

It is, of course, true that in certain specific instances, this mix of solutions has not yet proved effective. No one denies the potentially deleterious effects of DDT on all kinds of animal organisms or its dubious value as an ingredient in mother's milk. Yet, so far, no effective alternative has been found for dealing with the problem of malaria. Malarial spraying accounts, it is true, for only about 15 per cent of the general use of DDT. This may be an irreducible minimum in areas where the disease persists. But it is all the more vital that for other uses—insecticides, the control of weeds—a much more balanced ecological mix should be sought and any massive all-purpose use of DDT phased out.

Farms and Jobs

One of the elements in the right balance of response to the problems of the Green Revolution concerns the use of what is, in most of Asia and in some regions in all developing lands, the most plentiful resource—and that is manpower. It is a matter not only of reducing unemployment and underemployment—essential as that is—but of using manpower in ways that are ecologically constructive and less burdensome to the environment. For example, one of the most useful elements in proper conservation of the soil could be a rational use of abundant manpower in the development of rural public works. Manpower is needed for the work of replanting, particularly if the degradations have taken place in vulnerable watersheds on which dams have been built. In many developing countries, forest concessions to loggers are so short that loggers are encouraged to practice a "cut and quit" policy. Programs of reforestation require a variety of skills—surveyors to plot the contours, soil scientists to determine the most useful tree species, foresters to advise on the care and maintenance of infant forests. But a great deal of the daily work could be provided by the local people. They could also be employed in the terracing of steep slopes for agricultural usage. Many of them only have work at the time of harvest and, even if double cropping is beginning to increase their opportunities in the wake of assured water and the new hybrids, they could still fill in some months of the year with useful works of environmental protection. It seems clear that this type of rural public works on a massive scale is one of the secrets of the ability of China to continue to feed its vast and growing population in areas where, traditionally, the threat of flood and drought has been particularly acute.

Rural mobilization could also be utilized for other aspects of a balanced ecological package deal in agriculture. Before many of the chemical insecticides appeared on the scene, a variety of pests were kept under control by hand picking or by wrapping and covering vulnerable vegetables and fruit. The careful weeding carried out by Japanese farmers was a central factor in the very high yields secured in the early days of Japan's agricultural modernization. Composting in deep pits, judicious use of night soil, building village latrines are all methods by which rural labor can be used to keep humus in the soil and prevent human wastes from contaminating the local water supply. And since winter warmth and extra energy can improve the human environment, however simple it may be, village plantations for firewood and even for the working of small wood-fed generators for local light and tube wells can begin to build a more attractive environment in the kind of settlements in which, until the twenty-first century, the majority of mankind will continue to live.

But the Green Revolution, for all its promise, will not necessarily lead to a balanced and ecologically satisfactory use of human and natural resources. The inputs—of improved seed, of fertilizer, of controlled water supplies—all require an increased investment of capital. If cooperative funds or public investment are lacking and the capital infrastructure of the new farming is left almost solely to individual initiative, the risk arises that the gains from the new agriculture will be concentrated in too few hands. The "big" farmer—the 50-acre man in Asia, the 10,000-acre man in Latin America—can hope for a larger return on his investment if he consolidates holdings, terminates tenancies, heavily mechanizes his operations, and turns loose the subsistence farmers or sharecroppers who scraped a living from his land before fertilizers and new seeds and insecticides gave him both the chance of higher returns and the need to secure them in order to cover capital costs.

In the Indian subcontinent, there are signs of this radical social change going forward on the farms. Between 1965 and 1970, the number of tractors in use in India doubled to 100,000 and it seems that at least half the area now planted to high-yielding wheat is being mechanized. Estimates from the Punjab suggest a 50 per cent decline in the need for labor per acre. Another estimate, this time for Pakistan, puts at 700,000 the number of farm tenants who will become landless and workless in the next fifteen years if mechanization goes forward with its present momentum. In many areas where feudal land tenure is still the rule, the engrossment of the gains by a small elite and the dispossession of the small farmers will be an inevitable and damaging consequence for the social and institutional framework of the farming system.

Any such large substitutions of scarce and concentrated capital for abundant labor are economically and ecologically unsound for a variety of reasons. In the first place, the small owner, working with his own labor on a family holding, has been shown in a wide variety of developing countries—India, Brazil, Kenya, Colombia—to produce more per acre than the bigger estate. Some of the highest yields are to be found in countries where acre limitations are strictly enforced. This productivity is secured not by heavy machines which drink gasoline and can easily damage fragile soils but by hand work with light equipment which is, by definition, less prone to generate ecological risks. Fertilizers and pesticides are less lavishly used, humus and animal wastes more carefully husbanded. Greater personal care keeps terraces in trim, shade trees planted, gullies forested. And earnings are not spent, as is so often the case in semifeudal economies, on acquiring more land for extensive use, thus pushing up land prices and driving more working farmers away from the soil. Nor are they withdrawn altogether from the rural economy by the development of Western standards of conspicuous consumption or an overaffection for numbered accounts in Swiss banks.

At this point, the imbalance developing on the farms begins to create larger instabilities. If the dispossessed are not to work on the land, if their labor is not to be productively used to raise the food, clear the irrigation ditches, level the fields, replant the watersheds, where *is* it to be absorbed? As we have seen, there is already a vast migration of underemployed and landless people from farms to cities produced simply by the growth of population which farming cannot absorb. If at this point labor-intensive methods of farming are abandoned, the great tractors move in, massive fertilizer inputs are introduced, and the monoculture sprayed indiscriminately by helicopter, it is not only the balance of nature that will be endangered. It is the balance of the villages, the balance of the cities, the balance of the whole social order. Towns already grow twice as fast as population. Cities grow twice as fast again. If the rate becomes even more relentless, the environmental horrors of the great urban areas could become a key element in anarchy and breakdown.

What can be done to prevent such an outcome? The answers all involve an element of genuine ecological balance. First, the Green Revolution needs to proceed within a social framework of land reform and popular participation which provides maximum employment and the optimum distribution of gains from the new productivity. Next, its highly sophisticated methods need to be, as it were, "encased" in a context of extension services, agricultural research, farmers' training centers, adult literacy, and strong supervision, which, for instance, permitted Denmark in the nineteenth century to develop Europe's most productive agriculture round small farms, cooperatives, and peoples' high schools.

Next, this framework of expertise needs to be profoundly rooted in the environmental realities of local soils, climates, and plant varieties and take into account all the traditional wisdom that practical farming has developed over the millennia. The kind of ecological mix that is required is not one that can be brought in ready made from highly mechanized, temperate farm systems. It is the combination of modern science with local inventiveness and local responsibility that is ultimately at the core of the only really effective and sustainable ecological balance.

Last of all, it is no use pretending that the whole process of developing a productive and ecologically stable agriculture capable of feeding over 5.5 billion people in the developing lands by the year 2000 will be cheap. The kind of investment needed simply to provide seeds, fertilizers, and water supplies for the Green Revolution will, according to the FAO, demand at the very minimum $50 billion in cumulative investment by 1985. To add the essential infrastructure—of scientific research and personnel, of transport and market structures, of cooperatives and extension services could easily triple the sum. Of the kind of trained people and deployable capital required for such an effort there is at present little sign. Nothing on the horizon resembles the vast availability of free temperate land in the New World and Australasia which carried the Atlantic world over its early crisis of modernization in the 1840s and thereafter provided the safety valve for massive migrations out of industrializing Europe. The risk is that the immediate gains of hybrids and fertilizers, of large dams and assured water supplies will be introduced without bringing with them the needed supporting scientific services, the feedback, the monitoring, the technical marketing expertise. To this is added the danger that too much of the change will be financed by too small a range of investors with increasing damage to the balance of the social environment. But these risks—of inadequate local adaptation, of high technology introduced ahead of a suitable technical and scientific context, of capital scarcity, of too narrow a distribution of rewards and hence deepening pressures on the social environment—are not confined to the farming sector. They are social diseconomies that appear again in the development of industry and in the chaotic growth of the developing cities.

Industry: Employment and Pollution

When we turn to the industrial sector, growth and modernization of developing countries in the last thirty years have been faster, historically, than the similar stage of development in the Atlantic world in the early nineteenth century. An overall growth rate of 5 per cent was achieved throughout the late fifties and the sixties compared with a European average of about 3 per cent. And, until the recent startling increases in agricultural productivity, the bulk of the growth came from the priority given to high investment in power, transport, and the whole range of mechanical equipment in industry. The methods, the technologies, the processes were more or less taken over from developed models and represent the largest and speediest effort of forced-draught modernization since the Soviet Union's Five-Year Plans in the 1930s and the most widespread and sudden intervention of man's works in pretechnical societies ever experienced in history.

The results of the effort are now very generally admitted to be mixed. The achievements are genuine enough. Power supplies have nearly doubled. Continents whose entire transport systems were geared to the transfer of raw materials to ports enroute to developed lands have started to build up their internal links. In many countries, rates of industrial growth, admittedly from a very low base, have exceeded 12 per cent a year. The overall average has been a promising 7 per cent. But there are a number of storm signals appearing, pointing to difficulties which are much more complex than those which, for instance, caused a dangerous slowdown in early Atlantic development in the 1840s.

The first concerns capital. Much of the growth in the 1950s and 1960s was done—like America's between 1800 and 1850—by borrowing heavily from abroad. The external public debt of developing countries grew from $20 billion in 1961 to nearly $50 billion in 1968. If one adds debts incurred through foreign private investment, some countries, notably in Latin America, now need to set aside a quarter of their earnings of foreign exchange each year simply to service their debts.

This external difficulty is matched, in some countries, by an internal one. Too much of the newly produced wealth was engrossed by groups who did not reinvest it in the most productive way. In some countries, these groups were public officials bent on beautifying capital cities or building prestige steel mills in countries without iron or power. In others, private groups reinvested in real estate or exported their gains, thereby adding to the pressures on critically scarce foreign exchange.

The second set of difficulties concerns trade. Primary products still make up two-thirds of the exports of developing countries and only the oil-producing lands can expect to make very large and steady gains from such exports. Elsewhere the outlook is not promising. Natural fibers and rubber take the full brunt of competition from the world's synthetic products. Foodstuffs are heavily protected in developed lands. Tropical beverages are taxed in Europe. Although mineral prices jump up and down from feast to famine, any sustained rise in price encourages industrial communities to substitute other materials. Increases in the local fabricating of goods often incur a tripling and more of the developed countries' tariffs against them. These and other restrictions radically affect the developing countries' ability to secure what is still a critical part of their capital input—foreign exchange.

The third difficulty arises as a result of the nature of much of the technology introduced during the all-out effort of industrialization. It is broadly true to say that it has simply been taken over, often under foreign management, from the already developed lands. As such, it is based upon a mix of factors of production which does not necessarily match local conditions. The whole trend of research into new technology in developed lands is based upon highly trained scientific skills and maximum use of labor-saving technology. This is the precise opposite of local needs in developing lands. There capital is scarce, skills are scarce, labor is abundant. The result is twofold. The overall increase in manufacturing has provided jobs for not much more than a fifth of a labor force that rose by 200 million between 1950 and 1970. At the same time, population increase, inadequate employment in farming, and now the unsatisfactory rates of manpower absorption in industry are steadily increasing the rates of unemployment in the cities.

Now these difficulties—inadequate supplies of capital, both externally and internally, blocked trade, inappropriate technologies, rising unemployment, and pressure on the cities—are all ultimately and immediately concerned with the external diseconomies of rising social pressure. They have begun to stir up, particularly in relation to urban areas, profoundly disturbing questions about social justice and the quality of life. High consumption for the few, rising misery among the marginal men, restless countrysides, and cities which offer little hope of sustained and satisfactory employment for at least 25 to 30 per cent of the people —are these the recipe for a decent society? The question confronting developing countries when they examine the issues of resource use, pollution, and human settlement is thus the degree to which they are superimposed upon an already highly unstable economic and social situation in which further drives for economic growth, carried on regardless of social diseconomies, look like making the total situation even worse.

On the other hand, being a latecomer to the technical revolution is not all loss. In the first place, developing nations can now, in principle, enjoy more flexibility in industrial strategy. They can avoid the chief geographically limiting factor of the first Industrial Revolution—its commitment to steam, coal, and iron ore. Over the last century, the invention of electricity, the growth of alloys and plastics, the steady interchangeability of resources opens up a far wider range of location and variety for industrial activity and lessens an almost literally iron dependence upon a particular mix of resources. Provided energy is available, nations without iron ore can turn to bauxite for aluminum. Nations with oil and natural gas can solve both their energy and their resource equations and such nations include some in Southeast Asia, parts of west Africa and Latin America, and, of course, abundantly, the Middle East.

Coal can be gassified or turned to oil and fulfill a similar multiplicity of purposes. These reserves again are reasonably well dispersed.

Other energy supplies are also available. In Latin America and Africa there are still very large untapped resources of hydroelectric power— Latin America has 30 per cent and Africa 40 per cent of the world's reserves. Indeed, the Inga Rapids on the Congo River equal in potential the entire installed capacity of Western Europe. As research increases the possibility of transmitting high voltages over larger distances, regional grids offer the promise of decentralized power. And even where fossil fuels are relatively scarce and uncompensated by abundant reserves of hydroelectricity, nuclear energy is becoming a possibility which definitely is *not* rooted to particular sites and locations.

One of the great shortages in many developing lands—and in parts of the developed world—is water. The possibility of combining the generation of nuclear power with the desalinization of sea water affords the promise of creating agro-industrial complexes where they never could have existed before.

The issue is wider than one of increased choice. Industrialization is, after all, a fairly hit-and-miss affair. Some techniques work. Some demand more in capital than they ever achieve in income. They go on to fail in a string of bankruptcies or reduced consumption. In theory, it could be very useful for two-thirds of humanity to have the pioneers make some of the worst mistakes for them. They can avoid the errors and reap some of the rewards. Nowhere perhaps could these advantages be clearer than in the location and layout of human settlements. It is too late to cancel all the errors already made, but developing nations are not so profoundly embedded in the concrete and iron of an already urbanized order. Over the next decades, urban populations in developing societies will grow by as much as the entire present urban population in developed lands. There must and will be hundreds of new settlements. For them the master invention of the modern world, the generation of electricity, provides a freedom of location unknown in the early days. In theory, they could use decentralized industry, based on regional power grids, to offset the sucking pull of their big, inherited "export cities." They could study the whole variety of attempts now being made in developed lands to counteract the pollutive and social evils of vast urban conglomerations and to produce a more civilized and ordered use of their most useful and finally unexpandable resource—the resource of space.

This last need is particularly urgent because, in the great traditional cities of the developing world, the pollutive horrors of early industrialization have already appeared. Kanpur in India, with its long-established textile industry, very much resembles Manchester in 1860 save that it is already more than six times the size and the effluents of industry are estimated to equal those of another whole city of 1.5 million. The Damodar, running through India's major industrial area in West Bengal, carries effluents which equal the sewage of a city of a million people; yet two-thirds of the industries have no treatment systems of any kind and include in their pollutants such killers as cyanide. In Bangkok, where three million people and a growing industrial complex discharge all their effluents untreated into the Chao Phraya River and into the city's canals, experts estimate these water systems to be well on the way to irreversible degradation. The Pasig River, which serves Manila, is within sight of total contamination since, in periods of low water, it already uses up four-fifths of its carrying capacity for effluents—40,000 pounds of organic material a day, between a quarter and a half of it from industry.

However, as is the case with settlements, the major part of growth lies in the future and it is in the development of new industrial complexes that the most modern antipollutive technologies may become important.

It must be said, however, that many of the disadvantages we have discussed exist largely in theory. There is an element of Cloud Cuckoo Land in discussing what, in the best of all possible worlds, could be done in view of so many evident blockages and restraints. We have to ask, much more modestly, what can be done in the next critical two or three decades? How can resources be mobilized with least social and environmental diseconomies? What strategies best match the particular need? Above all, how can the worst of all pollutions, the daily grind of hopeless poverty on the fringe of farm and city be countered by a growth of opportunity and work, of shelter, of health, of hope?

Mobilizing Resources

For convenience we will look at the issue first from the point of view of what the local community can accomplish and then the possible contribution of our wider planetary society. But the distinction is fairly arbitrary since the economies of the developing world are still overwhelmingly dependent upon the flows of trade and investment which encompass the world's entire economic system. Even areas as vast and skilled as China are still, for instance, massively dependent on the world's wheat trade. In most ex-colonial areas, economic dependence continues to be almost as great as in the times of direct political control.

We began with national governments and the possibilities of action open to them. In resource use, they face three difficulties. The first is quite simply that many of them still do not know what they have. They can hardly plan for the optimum use of their resources before they have some kind of inventory of their soils, mountains, vegetation, and possible dam sites. Nor can they mobilize them without some reference to existing use. The United Nations Development Program has put special emphasis on this essential need to compile resource inventories. The government of Malaysia recently completed a very useful survey in west Malaysia. A national land survey was conducted, region by region, to establish the quality and hence the use of the nation's limited patrimony of land. New techniques of aerial photography provided reasonable detailed knowledge of the various types of soil, contours, geological areas, and forest cover. These could be related to existing uses—towns, cities, traffic networks. Thus the exercise permitted classification of the land to its capacity and potentialities for optimum use. In the process it was, for instance, discovered that some farming land would be better in forest reserves and some forest was covering prime arable land.

The report has been criticized for allowing such immediate needs as the location of minerals to outweigh dangerously the need for land set aside for conservation. This is surely specious. The first need for the rising millions in developing lands is precisely to discover what are the possible *economic* bases for expanding wealth and, above all, employment. New minerals, not sold abroad but developed through local processing and fabricating, are a clear bonus and to give them priority makes obvious sense.

Yet it is also true that conservation in the fullest sense covers all nature's incredible variety of vegetation, of forest masses, and animal species. In many developing areas, particularly in Africa, the only hope of safeguarding some of the earth's diminishing numbers of precious plants and animals in their natural habitats is to make them a part of the nations' long-term planning. In the words of a declaration issued at Arusha in 1961 after a conference of the International Union for the Conservation of Nature in what was shortly to become Tanzania:

The survival of our wildlife is a matter of grave concern to all of us in Africa. These wild creatures amid the wild places they inhabit are not only important as a source of wonder and inspiration but are an integral part of our national resources and of our future livelihood and well-being.

This last point—resources and livelihood—refers to two possibilities. The first is the possibility that in some developing regions wild animals, properly herded, managed, and "cropped," may provide a better basis for local diets than the introduction of exotic species.

The second is the extent to which properly maintained and serviced national parks can become heavy earners of foreign exchange through the growth of tourism. The passionate appeal of nature's unspoiled majesty is a reminder to urban man of the values he suppresses in his own imagination and spirit if the spread of modernization implies nothing but the growth of ugly cities, scarred countrysides, and a wildlife heritage virtually reduced to cockroaches and rats. Every developing land has some place of splendor and beauty, some species of unique interest to preserve for its own citizens' recreation and for the enjoyment of all the planet's increasingly mobile earthlings. The lesson of the developed nations is that such areas of beauty and recreation are not preserved by default nor by unregulated commercialism. There must be proper safeguards, firmly enforced. And for this, the first step is to establish where they are to be found and what natural systems they encompass. This emphasis on the natural heritage is, however, only one part of the nation's patrimony and no one will thank a team of experts who plot the best stands of balsam firs but leave out the lead and tin.

Even where inventories have been taken, governments in developing lands face a second problem—the relative value of various resources. Here they confront precisely the same difficulties that we have already studied in relation to supplies available for developed lands. A reef or a lode which has no value at one level of world prices or uses can become almost invaluable at another. A classic example is the uranium discovered in the hitherto useless mine tailings of South Africa's gold mines— a stroke of fortune which is not repeated in equivalent tailings in Ghana's Ashanti gold fields. But in this whole imprecise area of resource use and price, the developing countries do labor under a number of disadvantages.

They may find the whole process more expensive. The developed nations, among all the other advantages of getting in first, have made a killing with the more accessible and exploitable minerals. In some areas, Ghana, for instance, they removed nearly all the manganese with no more local costs than subsistence wages. Billions of barrels of oil left the Middle East at rock-bottom prices before the postwar mood of decolonization spread to the Persian Gulf. The first deal for any raw material extraction in which foreign investors and the local government divided profits on a 50/50 basis is barely twenty years old. It follows that there is an underlying likelihood that latecomers may find their mineral inputs more expensive than did the pioneers. Moreover, in some cases, they may find them exhausted. It has been estimated that if rapid industrial development were to bring all the world's peoples up to today's average developed standards, most minerals would have to be produced at a level at least five times higher than today. On such a calculation, sources of zinc, copper, lead, and tin could be out of circulation in a couple of decades. A lot of other metals would probably become much more expensive. Developing countries very naturally want to know by how much such contingencies may impede development, or at the least, add substantially to its cost.

Another difficulty is the continuing dependence upon developed markets for the profitable sale of raw materials. Some producers have started the process of setting up countervailing power by joining with major consumers in commodity agreements designed to stabilize prices and supplies. The oil and coffee producers have had a genuine impact. The problem is that too high a price for a natural material invites rapid substitution, and modern scientific technology has a vast new range of possible substitutes to offer.

The surest way of lessening the old dependence may be consciously to detach one's country from existing international trading patterns and to pursue a separate path of development. Yet, for any area less vast and well-endowed than the United States, the Soviet Union, or China, this may not be a viable path. It also brings us to the third difficulty—the sheer expense, the scale of investment, the abstention from consumption, the underlying austerity which rapid, effective development over the next few decades seems certain to impose.

Once again, we are up against the shortage of capital. If we talk cheerfully of the new technologies based upon synthetics, of nuclear energy which can be developed almost anywhere, of industries making full use of new techniques of recycling and in the process putting an end to most forms of pollution, we are talking of the technologies which demand the highest inputs of capital and the highest development of human scientific and technical skills. The American Council for Environmental Quality speaks of $4 to $5 billion a year to be spent by industry on new antipollution measures. Other estimates suggest a 5 to 10 per cent increase in costs to clean up past pollution and stay clean in future. But developing countries are hard put to it to raise the capital even for existing, cheaper, though more pollutive, technologies and energy systems. Nor have they the range of scientific skills to jump, at one stride, into the world of polymers and lasers and molecular welding. No doubt it would be a very desirable jump. But the capital is lacking and one thing only is truly abundant—vast· and rising armies of labor, much of it unskilled.

It is for this reason that if at this stage of development governments have to make a choice between providing more employment or keeping clearer air and water, there is every likelihood that the decision will be made for employment. Even in developed countries, organized labor tends to make this choice at the least sign of increasing unemployment. Developing governments will argue that air and water can hopefully be cleaned up later when new and possibly cheaper techniques are available. In the short run, human beings must not starve. Besides, they might add, many rivers and airsheds are not yet in any way overburdened. Is it not both environmental and economic sense to treat them as free goods? These are not abstract problems. In the wake of the Japanese government's stern and sudden resolve to end the deepening degradation of their own environment, a number of firms using highly pollutive technology are beginning to move to other more lenient areas. The concept of "pollution havens" appears for the first time in the international economy.

It is easy to argue that no self-respecting government would wish to expose its people to an extra risk of pollution and especially from the Japanese, who are now courageously denouncing the cost of allowing growth-above-all-things· to destroy their natural environment. But if a government is coping with an annual 2 per cent increase in its labor force and an overwhelming need for urban employment, there are real dilemmas to be faced in having to decide whether to put either pollution or employment ahead of the other.

However, given the existence of still-undeveloped land and water systems in many countries, a calculus can be made, balancing existing degrees of pollution against the burden that might be added by new industries. One should not invite a smelter into an industrial area whose effluents have already finished off most of the local estaurine fishing. But it could be sited on another part of the coast as a phased element in a policy of industrial decentralization. Such careful calculations of pollutive dispersion might act as counterforces to the compulsive tendency of industries to gather in big cities where labor force, markets, and services are already in existence.

At the same time governments can keep a keen eye on the development of antipollutive design and technology in already developed countries and make certain that, as they become both more efficient and less expensive, all new investments, local or foreign, are compelled by law or contract to include them. Since, in most developing countries, industry does not yet account for a quarter of national product or employ more than a fifth of the workers, part of the industrial growth ahead could, with careful planning, leapfrog the most primitive pollutive stages of industrialization and start with newer technologies installed as integral parts of the original design.

Such a strategy of careful attention both to the degrees of pollution existing in various parts of the country and to the possible adaptation of new antipollutive technologies to local use, is, in one sense, simply another aspect of the need for adaptive strategies. In the critical field of employment, too, governments can aim at the optimum mix of appropriate modernization together with the preservation of traditional labor-intensive activities. Large-scale electricity grids, ultimately powered by nuclear energy, do not exclude small diesel or wood-fueled generators in the villages. Precision instruments in one part of the factory do not mean that most of the moving or packaging of goods cannot be done by hand. Electric sensors may reject whole sheets of plywood in America. In Korea, knots are carefully removed and repaired by hand. In a sense, a policy based upon flexible concepts of technology and employment repeats the ideal of ecological balance. In every web of work and food there are intricate varieties of size and skill and it is precisely the overconcentration on the former, usually the largest and toughest, that creates dangerous instabilities in the whole environment. The automated steel mill in an unemployed countryside is as ecologically unbalanced as an overfed carp in an otherwise empty pool.

But all these suggestions—national inventories of resources, a rational mix of up-to-date technologies with labor-intensive forms of work, a determined effort of industrial and urban decentralization to areas less environmentally overloaded with existing industry and pollution, all depend upon one precondition. Governments must be determined enough and have command of sufficient resources to undertake large-scale public operations. It is costly to mount surveys, to finance desperately needed research, or to design pilot projects for better technological mixes and more labor-intensive methods of work. It is exceedingly expensive to provide the massive governmental inducements or investments required for a dispersal of economic and urban expansion to regional poles of growth. A government without authority or funds cannot begin to cope either with the promise or with the risks of the new technologies and ecological imperatives.

However, if governments in developing lands do not permit the concentration of the rewards of growth in too few hands—either in industrial and urban affairs or in agriculture—if they safeguard themselves through taxation and public savings against any shortfall in their own desperately needed supplies of public capital, then there are a wide variety of new techniques and opportunities which they can successfully adapt to their own local conditions.

And provided these principles of policy are observed, it may well be found that, at this stage of their modernization, there is no single policy that deals more adequately with full resource use, an abatement of pollution, and even the search for more labor-intensive activities than a planned and purposive strategy for human settlements based upon relocation and decentralization.

12 PROBLEMS OF HUMAN SETTLEMENTS

THE FIRST NEED is to take the strain off existing cities. It is here, in the growing number of urban centers in developing lands with more than two million inhabitants, that the worst human environments in the world are to be found. In India, for instance, despite the higher per capita income in the cities, the number of families living in one room is only 34 per cent in the country, 44 per cent in all urban areas, 67 per cent in the four largest cities. In Calcutta the ratio reached 79 per cent and still leaves uncounted the thousands for whom the pavement is the only home.

Great city figures for sanitation and sewage are usually horrendous. The cases of Bangkok and Manila can be matched by Djakarta. It has a population of nearly five million but since it remains in essence a vast agglomeration of rural villages it has no sewerage, no reliable drinking water, few transport links, and no policy whatever for the location of markets or industries. In Latin America, in spite of a well-established urban tradition, one can find similar levels of preindustrial sanitation. In Brazil only 45 per cent of the municipalities have reliable water supplies, only 34 per cent sewage systems. In Chile only 29 per cent have sewage systems. Yet some of their cities have passed the two million mark.

Above all, the vast migrations of people from the countryside, causing some of the greatest cities to grow by 8 per cent a year, have surrounded them by rings of shantytowns in which shacks made from gasoline tins, old automobile tires, and packing cases give a miserable cover to migrants who can still arrive—as in Rio de Janeiro—at a rate of 5000 a week and look like producing megalopolises of ten to twelve

URBAN POPULATIONS OF INDUSTRIALLY DEVELOPED
AND DEVELOPING COUNTRIES

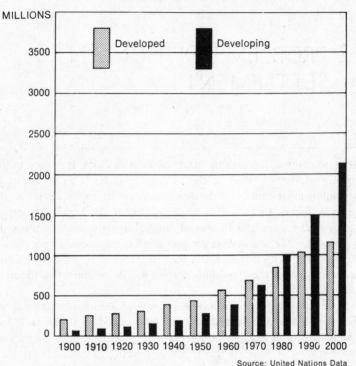

Source: United Nations Data

million, three-quarters in dire poverty, only ten years from now.

In confronting pollutions and miseries on such a scale, governments
may be tempted to throw in their hands. But there are a number of
policies, mutually supportive and reinforcing, that can be undertaken,
provided one preliminary step is taken—and that is not to repeat the
ridiculous complacency of market economies and suppose that com-
modious, convenient, and beautiful cities with a proper mix of classes
and an underlying sense of social balance and *esprit de corps* can be
produced by the unregulated operations of a speculative land market. In
few areas have single-thrust economics, transferring the profits created

by social needs to private owners, so radically influenced and often distorted the functioning of the whole community. If with all their wealth, Western societies have produced so many inconvenient and ugly cities—not to speak, in some of them, of flagrant injustices—developing societies, short of capital of every kind, will fare infinitely worse if they allow their urban land values and building costs to rise to the point where they are unable to afford an urban policy, however well they may have designed it in the first place. In India, for instance, the element of land costs in housing in big cities is already three times higher than in smaller towns and this is a major reason for the increasing degradation of big-city shelter.

The first principle is therefore a policy for land use, for the purchase or close control of urban land, and either the prevention or the securing for public purposes of speculative gains from rising land values. Without these, there is simply no conceivable way in which developing megalopolises will improve for the majority of their inhabitants. On the contrary, they are certain to grow worse.

But given strong authority in the urban field, developing countries can elaborate on a number of policies which have not been without effect in developed lands. One of the most important is the concept of building up countercenters or regional "poles of growth" to take the strain off center cities and to dam up some of the migratory flood which still moves inexorably toward them. Such planned centers must, of course, as in the Malaysian Plan, take fully into account the realities of local resources, networks, existing sites, and so forth. They must also reflect a dynamic understanding of the country's possibilities of development. But provided local plans are evolved with a full sense of variety and opportunity, they may reinforce the chance that economic influences, as the seventies open, are making for a more rational decentralization of industrial and hence of urban growth.

We have to remember how many of the older cities in developing countries are export cities that grew up as entrepôts of exchange between a single modernized export sector shipping out the primary products of mines and plantations and foreign businesses shipping back manufactured imports. Such cities have been, in a sense, extensions of the developed economy, not vital centers of local activity. So long as the chief economic forces in the developing world were mobilized behind the old unbalanced and dependent types of world trade, the functions of the export cities were economically too specialized to expand local business opportunity—and hence employment—when the great internal migrations began after the Second World War. They were magnets for the growing rural surplus. They were homes—of a sort—for the rising tide of urban babies. But they were *not* capable of extending their economic base to absorb fully a work force growing by 2 per cent a year.

In the seventies, however, two forces are modifying the relative weight enjoyed by traditional export sectors and cities. The first is industrialization based upon the local manufacturing of formerly imported goods. Whatever critics may say—often rightly—about its relative inefficiency and its overreliance on inappropriate capital-intensive technology, it has begun to broaden the internal industrial base. Wherever the processing of local materials is taking the place of their earlier export, it encourages the building of new towns sited near internal mineral reserves and pushes development back into the hinterland.

The link between agricultural modernization and regional, industrial, and urban development is quite clear. In fact, the two are very largely interdependent. Over the next two decades the Green Revolution is likely to modernize the inputs into farming—improved seed, fertilizers, power, machines—and dispatch the outputs to the commercial sector. All these complex inputs cannot be provided simply on a farm-size, retail basis. Villages are too small. Yet the big city is too distant. What is needed is the regional town as a wholesale and service center. In addition, credit operations require regional banks. Cooperatives need administration, extension services, and regional offices. Rural education must be backed up with secondary schools, training colleges, and research stations. Food processing can be performed efficiently by medium-scale enterprises. There is evidence to suggest that effective storage—a critical factor which could in some countries add up to 20 per cent to the grain harvest—can be arranged with the lowest overhead costs at the wholesale level in the regional town. Without this regional urban infrastructure, the agricultural revolution will itself be slowed down and the human avalanche into the major cities will not be slowed.

It is significant that, in the early stages of the Green Revolution, a seminar held in Kanpur to consider the regeneration of this run-down Indian textile center laid its whole emphasis on building around the city a network of smaller market centers, geared to intensive investment in the new agricultural technologies. These would provide the facilities, services, and processing plants that could help the farmers to reap some of the middleman gains from their increasing harvests. Their expanding income would, in turn, build lively markets for consumer goods manufactured locally by labor-intensive methods and for farm machines and fertilizer made in more elaborate plants in Kanpur itself.

In Kenya's planning, a kind of points system has been devised, in which each town is evaluated according to its suitability as a market center, its record of growth, its accessibility, its position in the country's grid of communications. The town with the highest score has the first chance of being picked for intensive development. In neighboring Tanzania, specific "poles of growth" to draw economic activity away from the capital, Dar es Salaam, have been written into the official plan.

These reactions spring from a new insight into development—the insight that in the context of creative planning, the needs of city and countryside are reciprocal. We confront once again the ecological principle that imbalance and insecurity result if only one thing is forced forward at a time. In the last twenty years, industry and infrastructure have received most of the emphasis. Now, with better ecological sense, we begin to realize that the developing world's potential agricultural revolution will not fully succeed unless the need for a decentralized network of urban markets and service centers is properly met.

And success in the agricultural revolution is a precondition of rational urbanization. Intermediate towns with lively economic opportunities rooted in productive farming could act as so many dams holding up the flood which now pours toward the great cities. It is, incidentally, in such towns that the health clinics, social education, and economic incentives necessary for responsible family planning would have most hope of beginning to reach the rural masses. The whole system would conform to the idea of countermagnets to the megalopolis—a concept that first evolved in developed lands but could be even more effective among the still unfinished urban patterns of the Third World.

If intermediate cities checked some of the flow of population, the task of upgrading existing metropolises would be less unmanageable. The new regional centers, built as in Holland's New Towns, on land bought and managed by public authorities, could provide industrial locations, office space, and areas for low-cost housing at reasonable rates. This would lessen the pressure on center cities and reduce not only the excuse but the temptation to go on piling up, for speculative gain, the kind of high-rise administrative buildings and high-rise tenements which make sense only where land is supremely scarce as in island cities like Singapore or Hong Kong.

At the same time, a check to the center city's sprawling growth will give a little time in which to reconsider two further aspects of present urban pressure. The first is the upgrading of the squatters' shantytowns. Experience, particularly in Latin America, has shown that the squatter groups have unsuspected vitality and initiative. Given help with the layout of the area, given assistance in drainage, water, and communal buildings, the squatters themselves often band together into improvement associations. Then they can transform a shantytown into a neighborhood or even an incipient suburb within a surprisingly short time. The provision of government loans to provide a part of the house (for instance, the roof loans in the Volta area in Ghana) encourages remarkable efforts of self-help. Moreover, evidence from all the developing continents suggests that the chance of a better home, made possible by state-sponsored mortgages, can be one of the most certain ways of mobilizing peoples' savings and making a frontal attack on a poor country's shortage of domestic capital.

The second issue concerns daily mobility. One of the greatest strains on squatter life is endless traveling in search of jobs, combined with wholly unreliable and ramshackle urban transport. Yet, at the same time, in the wealthy sections, the large, high-powered automobile is already adding twentieth-century air pollution to the still unsolved nineteenth-century search for effective sanitation. The contrast in transport is thus as brutal a symbol of social inequality as housing, schooling, health care, and everything else.

Yet the automobile is not yet out of control in developing countries. Before it becomes a premature mark of pseudomodernization, the decision is needed to compel the motorcar to carry its full environmental costs in cities—which will effectively banish it as a means of mass urban transport. Then the authorities can put in its place a variety of relatively cheap and, if necessary, subsidized modern bus services, supplemented by smaller units like the Hong Kong minibuses or the multiple-use taxis.

The decision of so many cities in North America to accept the necessity of improved and increased mass transit—in spite of immense wealth, car use, and consumer pressure—suggests that developing countries should take a very cool look at the costs of the automobile before allowing it to occupy a key place in their manufacturing patterns and exercise a runaway appeal to their consumers. The coming world shortage of petroleum also suggests the wisdom of waiting until rising costs compel developed countries to invent a less polluting but still usable form of private conveyance.

These are some of the strands in a policy for human settlements that would not repeat all the contradictions and shortcomings of developed cities. But vital as it is to take pressure off the centers, upgrade the neighborhoods, provide shelter and transport, the greatest need, in terms of human dignity and happiness, remains the search for work. Here, too, a policy of vigorous, decentralized urban growth can play its part, together with more productive agriculture, in increasing the supply of jobs. Once again we see the ecological balance between the farming and the industrial sectors. Modern agriculture can be labor-intensive, especially if double or triple cropping becomes possible as a result of better water management. Small farmers can, as in Japan, make a middle-class income. Laborers can get year-round work. Their increased income provides markets for manufacturing carried on in regional urban centers. And, in addition, the construction of these centers, together with rehabilitation in big cities, can act as a major stimulus to the construction industry. This is critical. The building trades are still the largest users and trainers of unskilled labor. Thousands upon thousands of migrants have first found their feet on the scaffolds of urban construction.

The need for new housing is so vast that planners have been daunted by it. In the developing continents, estimates of $12 billion a year needed for houses alone have caused governments and economists to avert their eyes from such horrific welfare expenditures. But if one looks at houses as a consumer durable for which people will mobilize their savings, if one reflects upon the degree to which a large building program stimulates local industries (building materials, timber, plumbing equipment, furniture), if one remembers that Britain's New Towns, set up with public capital under a statutory corporation, have paid a commercial return on the investment, it is possible to ask whether a vigorous policy of town building, coupled with the new opportunities in agriculture and the careful choice of labor-using technologies in processing and in light industry, might not prove one of the most effective means of countering worklessness and building up an urban environment in which it is worthwhile to live.

Yet, however many promising policies and strategies may appear to be open to developing countries in their search for a better human environment, a question mark hangs over the next decades. The inexorable mathematics of population increase, of city growth, of investment levels, and trade circuits makes it certain that the resources needed to feed, house, employ, and educate two and a half billion more people at anything like a reasonable living standard in the next thirty years will require a new and massive mobilization of local and international resources. How, otherwise, can cities which already hold one to two million people—a third of them squatters, a quarter of them without work —still prepare for a wave of two to three million more?

Nor does the fact that three and a half billion people will still be living in the developing countrysides in the year 2000 promise any better standards. In many lands, pressure on the land will have reached such proportions that tiny fragmented holdings, shortened fallowing, deforestation, and eroding soil will threaten the very survival of the whole farming system.

In the cities and on the land, only a massive and increasing investment of capital and skill can give governments and peoples time to evolve the kind of modernized, technological, high-productivity society based upon stabilized population, high investment, and skills which permit countries with the highest densities—like Holland or Switzerland—to enjoy among the highest per capita incomes.

Hope for developing peoples—and indeed for the whole international order—lies in the fact that such technologies are available, have worked, and are capable of creative adaptation. The difficulty, indeed the danger, lies in the degree to which they surpass the developing nations' own powers of mobilization, either of savings or of skills. They require a sustained strategy and effort on the part of the entire planetary system.

And yet that system, in the sense of a creative and responsible human community, is still only partially invented.

It is here that we begin to encounter the full paradox of the role of the nation in the modern world. The developing countries have no other basis of dignity, identity, and active political will upon which to base their passionate need for development. The nation-state has emerged as the master institution of the modern world. For developing nations in particular it is the profoundest symbol of their escape from servitude. But symbolism is not enough to cancel its inadequacies.

Many developing nations are too small for effective sovereignty. Virtually all of them lack the resources for the kind of growth and change they desperately require as a consequence of the convoluted, self-reinforcing nature of their problems. As technological latecomers to the tasks of development, they enjoy none of the free bonuses of land and minerals, the unimpeded movement of migration and settlement, the forceful growth of trade in markets kept open by colonial links and protection which first gave the present developed states their overwhelming predominance in economics and politics—a predominance they still enjoy. In fact, this inherited postcolonial order makes up the technical environment in which nations today have to modernize—the world of high (for them) inappropriate capital-intensive technology, of large, sophisticated competing corporations, of markets protected against labor-intensive goods, of research concentrated almost totally on the problems of the rich, of frontiers closed to the migration of the unskilled, of "braindrains" that attract the talents of the educated.

This techno-economic environment is as much a fact about planetary society as the air over the nations or the seas that wash their shores. We have to recognize that it is not only man who inhabits two worlds—the biosphere and the technosphere. Nations, as communities, display the same ambivalence. They cannot escape either. Even though man-made arrangements are more adjustable over time, they exercise in the short run as compelling and inescapable an influence as nature's systems of climate and oceans.

The difficulty is that nations are not yet ready to confront the facts of these larger technical and physical interdependences. Man's habits of self-government are almost as old as human history. Tribes, peoples, nations—coalescing in larger empires, falling apart again as old centers weakened and new ones arose—have all practiced separate decision-making. The habit of sovereignty has its roots, 40,000 years deep, in the autonomy of the hunting clan. In fact, the only universal claims to authority ever made in the past have been made by those whose sovereignty is now most violently rejected—the claimants to world empire, the imperialists who used a temporary political and economic predominance to dragoon other groups into subjection. That kind of "unity" is rejected in our day with unequaled vehemence by all nations, great and small. The world is committed to pluralism and decentralized decision-making —even though the theory may still exceed the practice in many areas.

How, then, is any perception of the essential unity and interdependence of both technosphere and biosphere to be combined with the acutely self-conscious separate sovereignty of more than 130 national governments? The first step is clearly to establish where these interdependences are already unequivocal, unavoidable facts.

Part Five: A Planetary Order

13 THE SHARED BIOSPHERE

Airs and Climates

NOWHERE IS THE vulnerability and interdependence of the total biosphere more evident than in the envelope of atmosphere upon which more and more of industrial man's activities are beginning to impinge. It is, of course, alien to our thinking that the firmament itself could be vulnerable to our intrusions. Yet there can be a useful corrective to this kind of thinking in a return, for a moment, to our knowledge of the alphabets of space and time—of the electromagnetic spectrum and the earth's evolution through the billennia. Let us recall the solar-shield effect of the earth's atmosphere. Over the ages, the general level of heat in the planet has been maintained with fair uniformity by a critical balance. Incoming solar radiation, coupled with the earth's own reabsorption of the heat it gives off itself, just about equals the amount of radiation that is blocked en route from the sun or sent from cloud and earth surfaces back into space.

Different parts of the planet are, obviously, warmed and cooled in different degrees and their interchanges—through the winds and air currents and the universal mediations of the ocean—make up the whole, totally interdependent climate of our planet. In the tropics, more heat is absorbed than in the highly reflective polar regions. The heat generated in the center tends to flow to the poles and their cooler airs are drawn back toward the center. The general effect is to mitigate extremes of temperature. But this relatively straightforward motion is immensely complicated by the spinning of the earth on its axis, by the massing together of land in some areas, of water in others, by high mountain ranges, and the distribution of rain forests and deserts. With so many variables, it is not surprising that local weather systems exhibit very large variations round expected norms. It is perhaps equally unsurprising that the whole global climate can itself undergo profound modifications.

During about 90 per cent of its stable existence, our planet appears to have had no ice at all at the poles. But we know from the evidence of geology that it has undergone some five or six periods of glaciation. We appear to be in the tail end of the latest—the Pleistocene Ice Age which lasted over a million years and brought the glaciers to the Mediterranean. Today, the ice has retreated, but is not quite back to normal. Yet so great is the immediate effect of the ice caps on our global climate that "normal"—in other words, no ice caps—could mean a catastrophically different topography, with some land masses under water and others indescribably hot.

Clearly man has had nothing to do with these vast climatic changes in the past. And from the scale of the energy systems involved, it would seem rational to suppose that he is not likely to affect them in the future. But here we encounter the other facet of our planetary life: the fragility of the balances through which the natural world that we know survives. In the field of climate, the sun's radiations, the earth's emissions, the universal influence of the oceans, and the impact of the ice are unquestionably vast and beyond any direct influence on the part of man. But the *balance* between incoming and outgoing radiation, the interplay of forces which preserves the average global level of temperature appear to be so even, so precise, that only the slightest shift in the energy balance could disrupt the whole system. It takes only the smallest movement at its fulcrum to swing a seesaw out of the horizontal. It may require only a very small percentage of change in the planet's balance of energy to modify average temperatures by 2°C. Downward, this is another ice age; upward, a return to an ice-free age. In either case, the effects are global and catastrophic.

Scientists are, therefore, turning their attention to the points at which human actions, however minuscule their effects may seem when set against the total scale of the planet's energy system, may nonetheless trigger off one of those small but fateful changes which alter the balance of the seesaw.

Among the enormous range of technological man's activities, three such points of leverage seem serious enough to arouse real concern. The first turns on the role of carbon dioxide in intercepting the earth's heat radiations and in transmitting them back to the earth, the so-called greenhouse effect. Its action is rather like the effect of car windows when the sun's rays enter the car and warm up the seats and fittings inside. But glass does not pass on heat. It lets the rays through but retains the heat within the interior of the car, which gets hotter and hotter. In the atmosphere, carbon dioxide can have the effect of glass. It can cut down the rate of surface cooling and we do not know whether its effects are reversible. In normal amounts—the .03 per cent of the total atmosphere —it plays a very small, though specialized, part in the earth's heating system. But there is evidence to suggest that over the last decade, the release of carbon dioxide into the atmosphere as a result of man's burning fossil fuels has been increasing by 0.2 per cent a year. We simply do not know where all the carbon dioxide produced in the biosphere year by year actually goes. Perhaps half is absorbed in the oceans and the metabolism of plants. But the increasing concentration in the air means that, at present rates of use, the earth's temperature could rise by 0.5°C by the year 2000.

But present rates may well increase. Excessive deforestation can reduce the rate of natural removal of carbon dioxide from the atmosphere through the action of leaves. At the same time ever greater amounts are being pumped into the atmosphere as industrialization goes forward. The energy demands of developed societies are still rising sharply. Projections of power demands in the developing world suggest even more precipitous increases. What would be the consequences of multiplying energy consumption in the developing nations to the levels obtaining in technological societies? We do not have to postulate the fantasy of three and a half billion cars on the planet to begin to wonder whether the sum of all likely fossil-fuel demands in the early decades of the next century might not greatly increase the emission of carbon dioxide into the atmosphere and by doing so bring up average surface temperature uncomfortably close to that rise of 2°C which might set in motion the long-term warming up of the planet.

The risk is increased by the possibility that a change of this kind could conceivably be working with and thus reinforcing an underlying global movement already taking place independently of man. Recently scientists have extracted long cores from the Greenland ice cap and constructed from their variations in freezing and melting a sort of profile of an ice age. The last two are found to resemble each other to a marked degree—the change from major glaciation back to ice-free conditions being marked by a series of remarkably similar wobbles back and forth between more freezing and less freezing. It is not therefore irrational to wonder whether a massive man-induced increase in the atmosphere's carbon dioxide, coinciding with one of nature's own warmings up, might not change a slight move at the center of the seesaw into a violent shifting of weight and the risk of major and unpredictable global consequences.

Another range of risks is incurred by industrial man's increasing emission of dust, soot, and gas, which combine with each other and with droplets of vapor to thicken up the atmosphere and increase the earth's cloud cover. The higher the altitude of these concentrations, the more lasting they appear to become. Particles which would vanish in a few weeks in the lower airs can last from one to three years in the high altitudes. There is already evidence that cirrus clouds are increasing along the most-used air routes in the northern hemisphere and that the earth's cloud cover as a whole is showing some signs of deepening. The difficulty is to know what effects such changes might have. If they effectively reduced the passage of the sun's radiation, they might lower the earth's temperatures. If, on the contrary, they reflect back the earth's own emissions of heat, they reinforce the greenhouse effect.

Scientists have some evidence to go on. In 1963, Mount Agung blew up, taking a sizable piece of Bali with it. Like the great explosion of Krakatoa in 1883, the Agung eruption filled the lower stratosphere with particles which dyed the sunsets with light-reflecting particles. The effects persisted for several years and had world-wide consequences within six months of the event. The band of the stratosphere lying above the equator was heated up by 6° to 7°C immediately after the eruption and remained 2° to 3° higher for several years. There is thus no doubt that gases and particles in the stratosphere do hang about, do have world-wide effects, and do raise the temperature. They may do other things—combine in unpredictable ways with each other under the direct, almost unshielded influence of the sun's radiation. A sort of photochemical effect like smog is conceivable. It has also been suggested that nitrates and sulfates from volcanoes—or supersonic exhausts—can unite with the critical supply of ozone and deplete the upper atmosphere of one of the essential elements in the planet's antiradiation shield.

It is, incidentally, this uncertainty over the cumulative effects of carbon dioxide, particulate matter, vapor, and gases in the atmosphere that has led some scientists to advocate caution in any massive development of supersonic transport. But virtually all scientists would agree on two propositions. Industrial man, by using the air as a giant sewer, can have profound and unforeseen effects on the earth's climate and thus the possible consequences will be borne not simply by the polluting agencies, but by the biosphere as a whole. From this follows the second point. We need far more knowledge, far more sophisticated simulation of climatic effects on giant computers, far more monitoring on a global basis, far more exact information on what we are actually doing in the atmosphere that the whole of mankind must share.

All these concerns with global air pollution lie beyond the effective protection of individual governments. It is no use one nation checking its energy use to keep the ice caps in place if no other government joins in. It is no use the developed nations suggesting lower energy use just at the moment when the developing nations see increased use as their only exit from the trap of poverty. The global interdependence of man's airs and climates is such that local decisions are simply inadequate. Even the sum of all local separate decisions, wisely made, may not be a sufficient safeguard and it would take a bold optimist to assume such general wisdom. Man's global interdependence begins to require, in these fields, a new capacity for global decision-making and global care. It requires coordinating powers for monitoring and research. It means new conventions to draw up ground rules to control emissions from aircraft and to assess supersonic experiments. It requires a new commitment to global responsibilities. Equally, it needs effective action among the nations to make responsibility a fact. And all these necessities—for more research, better monitoring, stricter control, and more global action—are simply reinforced when we turn to man's other universal environment—the world of the seas and oceans.

The Oceans

Only the fact that so much of the surface of our planet is composed of water makes it habitable. And, in the view of many marine biologists, the oceans are the most immediately threatened part of the biosphere.

It was in the oceans, after the secular downpour of the early rains, that life first began to stir, shielded by the waters from the sun's irresistible radiation. It was from the oceans that plants and animals emerged to colonize the land surface of the planet. It is the oceans today that provide the water vapor which, drawn up by the sun, falls upon the earth in harvest-bringing, life-sustaining rain. Ocean water is our planet's filtering system where all debris, both mineral and biological, is dissolved, decomposed, and transformed into life-supporting substances. It is the universal global sink, a vast septic tank from which clean water returns to man, beast, and plants by way of evaporation and precipitation. It is a major provider of the oxygen released by its phytoplankton for the benefit of all the species of land, air, and sea—breathing with lungs and gills. Without water's special qualities for holding heat, much of the earth would be uninhabitable. The oceans are the coolants of the tropics, the bringers of warm currents to cold regions, the universal moderators of temperature throughout the globe.

In more mundane ways, too, the oceans are indispensable to man. For good and ill, they first created the world-wide currents of seaborne trade which, since the fifteenth century, have steadily drawn our planet into a single economic system. And they still produce, for the globe's rising population, a vast harvest of indispensable protein. In 1969, 63 million metric tons of fish came from the sea. This is estimated to be only approximately one-fifth of the oceans' production. But most of it is deeper than can be fished economically with present equipment. Even so, the seas' fish harvest within its present rate of productivity could perhaps be tripled and an added 100 million tons, properly distributed, could provide some 20 million extra tons of protein annually to offset the deficiency which threatens so many of the planet's children over the next three decades.

Incidentally, one of the world economy's most unacceptable diversions of resources is that at least 50 per cent of the fish catch which today is converted to fish meal ends up feeding pigs and chickens in developed lands. If turned directly to human use, fish could make up part of a protein diet for the world's children at an annual cost of no more than $8 a child. Thirty years ago, only 10 per cent of the fish catch was diverted from human consumption. It is a sobering commentary on the humanity of our world economy that "developed" animal pets have the chance of a better diet than all too many "developing" babies.

As in the airs and climates, the ocean system must seem, at first thought, to be infinitely beyond the reach of man's puny influence. Whether we see it as the sacral cleanser of "Earth's human shores" or the cruel "widowmaker" of a million shipwrecks, hurricanes, and typhoons, it is restlessly powerful, serenely or menacingly indifferent to all the busy activities men seek to pursue in and under its buoyant, treacherous waves.

In fact, men are still under the strong influence of the medieval concept of an endless ocean. We all tend to feel that once a polluted river empties into the open sea, once we conduct city sewer systems far enough away from land, all industrial and urban discharge will disappear somehow into blue space beyond the horizon, as if we pipe it away from our own planet. We seem in this conception of the ocean to forget for a moment that the world is round and without edges. The first and only refuse man has ever disposed of outside his own biosphere is what was recently parked on the moon. Every ounce hitherto dumped or channeled into the sea, from the very morning of time until the modern age of general industrialization, has accumulated in one form or another inside the same landlocked sea, the lowest section of our biosphere and the only one with no outlet for refuse. Standing on the beach gazing toward the horizon where blue sea runs into blue space, we have not quite digested the message from the countless twentieth-century voyagers who, every year, cross the ocean from continent to continent, or the astronauts who see the whole planet from above, all giving their witness that the ocean has none of the infinitude we give it in our dreams.

Lake Erie is grossly polluted, to a level one would have thought impossible a few decades ago. Take ten Lake Eries and place them end to end and they will span the Atlantic Ocean. But the ocean is much deeper and spreads in all directions, we may reply. Yes, but Lake Erie has an ever-running outlet and hardly half a dozen major cities to pollute it. The ocean, however, daily receives and never returns the outlet from Lake Erie and thousands of other polluted lakes and rivers all over the world, and directly or indirectly the sewers and fallout from all the world's countless cities and all farmlands and all the industry. Instead of running into blue space, the oceans are landlocked and, if we go far enough in any direction, they are completely interlocked and share the rapidly accumulating pollution among them.

To this we must add that the waters most important to man are those most rapidly polluted: the layer nearest the surface and the coastal and estuarine zones. The bulk of plankton and other marine life dependent on photosynthesis is concentrated in an upper layer of ocean water no deeper than the great lakes. In fact, about 80 per cent of the world's fish catch is derived from waters less than 200 meters deep, which corresponds to half the depth of Lake Superior. Again, this concentration of biological life near the surface is further amplified next to land. An estimated 90 per cent of all marine life is concentrated above the continental shelves, which represent only about 10 per cent of the total ocean area. Plankton and fish indispensable for life on earth are thus concentrated in the ocean water most vulnerable to man's activities. There are fish outside and below these areas of high concentration, but in greatly reduced quantities, and also below the reach of nets; 1000 meters is fairly generally held to be the outside commercial limit.

Thus, while we have to readjust drastically our erroneous conception of an endless ocean invulnerable to the combined refuse of all men on earth, we must also face the fact that there are leverage points, areas near the surface and near the coasts in which human actions very soon can become sufficiently concerted to have lasting destructive effects. Even deep-sea species are dependent on the exposed coastal zones.

Estuaries and shorelines are almost without exception the spawning grounds for fish, some of which migrate over vast distances to swim back up the same rivers when spawning time returns. It is also at the level where land and sea meet that the most hopeful areas lie for any very great increase in "managed" fish cultures with the purposeful use of nutrients and heat to increase the fish catch in a systematic way. Shellfish, for instance, can be encouraged to double their output if the velocity of water round their beds can be increased. Hot water effluents from power stations, which damage some kinds of marine life, can be helpful to others. Shrimp harvests, for instance, appear to be responding to a managed use of heat.

Coastal waters are not only useful to man for food. They provide him with some of his favorite recreations. The annual day at the seaside was the supreme break from nineteenth-century dirt and toil. Two weeks on a packaged tour to Majorca or the Black Sea are the modern equivalents. Swimming, boating, skin diving, speedboating, fishing—the millions upon millions of tourists who get into motion as summer comes on usually end up with one toe in salt water. As leisure increases, the seas' importance for vacationers grows in the same degree.

And it is precisely the coasts and estuarine waters that are becoming steadily less usable for human purposes. Whatever awe man may feel at the majesty and magnificence of his planet's seascapes, he goes on treating the ocean as a sewer. In many lands there are heavy concentrations of population near the sea. A very large part of domestic sewage is simply dumped directly into the sea with minimal or no treatment. In addition, industry contributes a steady quota of heavy metals, inorganic materials, and, on occasion, radioactive waste. Rivers, too, add their effluents whenever, as is the common practice, they have been used as drains.

Rivers also bring down some of the run-off from fertilizers. And pesticides used in agriculture, above all, the chlorinated hydrocarbons, such as DDT, are carried out to the ocean, where, following the currents and concentrating as they go up the marine food chain, they affect animals even in the areas most remote from agricultural activity, like the polar regions. Chlorinated hydrocarbons sprayed by man as pesticides in farmlands turn up producing unhatchable eggs from the bald eagle, the peregrine falcon, and other species. They accumulate in the organs of polar bears. When twenty whales born and bred in the east Greenland current coming from the North Pole were recently harpooned for test purposes, six identifiable pesticides including DDT were found in the blubber of all of them.

Apart from the toxic influx of global range, other pollutants have roughly the same effect on coastal waters as they have in rivers and lakes. The nutrients from domestic sewage and agricultural wastes tend to overfertilize inshore waters. Blooms of marine plants become more frequent. A mixture of sewage and fertilizer run-off increased the bacteria count in New York Harbor ten times over between the late fifties and the late sixties. Where discharges take place into closed seas like the Baltic and the Mediterranean, there is real risk of producing permanently anaerobic conditions, in other words, such a lack of oxygen that only foul-smelling marsh plants and animal life can finally survive. When in 1971 many Italian resorts had to close their beaches for fear of widespread hepatitis, they only gave a preview of what might become the Mediterranean's universal condition after another decade of inadequate sewage treatment.

It is all too often the coasts which chiefly suffer from the pollutions caused by the drilling and transport of oil and from the even larger burdens brought to the seas and rivers in industrialized areas. Underwater drilling for oil is steadily increasing on the continental shelves. Even if, hitherto, disasters such as the massive oil leak in the Santa Barbara

Channel are rare, the expansion of underwater operations may increase the risk of more frequent mishaps in the future. So far, underwater drilling is largely carried out by techniques adapted from land-based procedures; blowouts do occur in land drills, but they do so with far less disastrous consequences than would be the case in coastal waters which could spread the pollutants with every force of wind and tide and current. At present, only 17 per cent of the world's oil supplies come from offshore drillings. By 1980, offshore production is expected to rise to 50 per cent of the oil produced from all sources in 1970. Such rapid expansion carries with it a danger of more frequent spills, more fouled beaches, more depletion of estuarine hatcheries, more birds and fishes dead in the slicks of oil.

It is also, as the disaster of the *Torrey Canyon* so visibly illustrated, in coastal waters that there is the greatest risk of groundings and collisions of tankers which release oil to the seas and beaches. Nearly all the major tanker routes lie inshore—the Persian Gulf, the Mediterranean, the western waters of Europe, the eastern waters of North America. The estimates of how much oil reaches the oceans from passing tankers, drilling rigs, and coastal installations have varied from 1 to 10 million tons annually but an expert group reporting to the Secretary General of the United Nations recently estimated that 2 million tons are reaching the oceans every year despite conventions regulating oil emissions

Another possible danger lies in the astonishing growth in the proposed size of tankers. Only five years ago, the average tanker was 12,000 to 13,000 tons. There are now four tankers in operation of 325,000 tons. Future plans point to monsters of 800,000 tons. A single disaster with one of these giants releasing all its oil would increase by 25 per cent the total pollution of the seas in that year. On the other hand, given the shortage of skilled crews and the fact that large tankers can absorb the cost of sophisticated oil-handling equipment, there is a case to be made for the possibility that larger tankers will be better serviced and so consequently safer than smaller ones.

Since 1964, major progress in restraining pollution from tanker operation has been achieved with the development of the "load on top" system. Tankers returning from their delivery point normally carry a ballast of sea water and clean their tanks and bilges at sea. The previous practice was for tankers nearing the end of their return voyage to discharge their ballast—oily wastes, sludge, and all—into the sea. Now, equipment on board and new facilities in port allow tanks to be cleaned at sea and the residues off-loaded on return to harbor. The ballast must still be discharged at sea, but with far less pollution—how much less depending on factors like the length of voyage, temperature, and efficiency of the crew. The adoption of this procedure by 80 per cent of the world's tanker fleet has reduced the amount of oil intentionally discharged into the ocean by 2 to 3 million tons a year. Extension of this practice to the rest of the world fleet could reduce the current 1 million tons deliberately discharged to a theoretical minimum of about 100,000 tons.

Oil pollution is more than a coastal problem. It is a consequence of the fact that man uses the whole oceans as a receptacle for unwanted materials and wastes. Here we confront a difficulty. We do not know how much is being poured into the oceans. We do not know how much they can stand. At least, in coastal areas, we can see—and smell—a clear and present danger and make some estimates of the costs of the damage and the expenditures needed to remedy them. The British taxpayer has some idea of how many millions he had to pay to clean up after the *Torrey Canyon*. Hoteliers can estimate their losses as beaches close in the wake of hepatitis. A balance sheet can be roughly drawn up between the decline in salmon and the rise in shrimps—although salmon's migratory habits make this a devious international calculus of profit and loss. But once we get away from the shores and from reasonably measurable and observable effects, we enter a world almost as dark as the ocean deeps.

Now it is perfectly true that nature itself uses the oceans as a dump. Every year the rivers of the world wash down into the oceans enormous quantities of minerals which are diluted or oxidized on the surface or sink to the bottom. Estimates made in the mid-sixties suggest that every year, nature, by its own unaided efforts, flushes some 25 million metric tons of iron into the seas and between 300,000 and 400,000 tons of manganese, copper, and zinc. In addition, lead and phosphorus each provide 180,000 tons and mercury 3,000 tons. The point about these last three elements is their toxic effect. Lead and mercury are lethal poisons; phosphorus contributes to algae blooms. To these natural flows we must now add the vast and accelerated run-off from modern technology. This inevitably increases greatly the flow of minerals to the final dump of the ocean. And these levels prevail at a time when only one-third of humanity has fully entered the industrial era. Extrapolations of comparable output and run-off to match rising world activity would mean a sobering amount of toxic substances in the oceans by the year 2000, simply as a likely by-product of the more or less casual disposal of industrial wastes.

And to this we must add the very far from casual dumping to which the oceans are exposed. Once again, we are in very great darkness. Every now and then a sensational fish kill reveals that old canisters full of mustard gas have at last eroded as a legacy of the First World War. The much-publicized dumping by the United States Army of containers of deadly nerve gas off the Bahamas in 1970 is remarkable chiefly for that publicity. One may wonder whether all military establishments resist the temptation to dump their poisons under the cloak of national security.

But this is not simply a military game. Some municipalities regularly dump their sewage in nearby stretches of international water. New York City has created a "dead sea" at the approaches to the harbor by dumping massive amounts of sewage sludge there for the last two decades.

As nuclear power generation increases, most nuclear nations are likely to drop into the oceans more of those hopefully secure stainless-steel containers loaded with radioactive wastes. But a lot of industrial dumping is much more casual. In 1970 only the alertness of the official maritime authorities in Norway exposed the practice of a number of European plastics manufacturers of dropping their most poisonous effluents in containers in the North Sea.

Nor is secrecy the only source of ignorance. We still know very little about the processes by which the oceans detoxify, evaporate, or absorb the massive wastes which flow into them. The various components of oil, for instance, differ greatly in durability and toxic effect. They differ too, in their response to temperature. Heat and oxidation are prime factors in dispersing oil's by-products. But in Arctic regions, frozen waters and sub-zero air may keep oil spills or leakages intact for half a century. We also know little about the deeper movements of currents in the oceans. It is not impossible that canisters of poisonous toxins disposed of in supposedly safe ocean trenches and deeps may, for all we know, be swirled across the ocean floors to end up, broken and leaking, on the rocky coasts of distant continents.

There is, in short, no escape from the underlying unity and interconnection of man's ocean world. Seas and oceans, like the airs above, mingle with each other, pass on each other's burdens, cleanse or poison each other, move in steady currents and unpredictable tempests to weave a seamless watery web. Their rains fall on the just and unjust. Their tides sweep every human shore. Sovereign governments may proclaim their sovereign national control over their own territories. But the airs bring in the acid rain. The oceans roll in the toxic substances. Pollution moves from continent to continent. And what is territorial water off Peru today becomes territorial water off Polynesia a few weeks hence. It is, above all, at the edge of the sea that the pretensions of sovereignty cease and the fact of a shared biosphere begins, more strongly with each passing decade, to assert its inescapable reality.

The governments' response to this imperative has been to move hesitantly into limited conventions and agreements with other nations to lessen the risks of degrading the oceans. The examples are beginning to multiply. Take, first of all, the question of pollution by oil. The nations have taken some preliminary steps to check it. The International Maritime Consultative Organization (IMCO) has been instrumental in working out a Convention for the Prevention of Pollution of the Sea by Oil, negotiated in 1954 and amended in 1962, limiting the rights of ships to discharge oil, prescribing safety features for vessels, and providing for rights of inspection. In 1958, two conventions, on the High Seas and the Continental Shelf, included provisions to keep oil from damaging the marine environment. Then in 1969, in the aftermath of the *Torrey Canyon* disaster, further conventions were negotiated—the Brussels Convention Relating to Intervention on the High Seas in Cases of Oil Pollution Casualties and the Convention on Civil Liability for Oil Pollution Damage.

In two related areas of possible pollution—the disposal of radioactive wastes and the dumping of toxic chemicals—the need for international action is also recognized and beginning to emerge. At present, the number of states conducting nuclear programs and facing a problem in the disposal of radioactive wastes is very small. The United States has recently discontinued the practice of disposing of radioactive wastes at sea. But by 1980 the world output of nuclear energy may be ten times larger than it is today and a dozen or more nations may have begun production of power from nuclear reactors. The International Atomic Energy Agency has started, under the general authority of the 1958 High Seas Convention, to lay down standards for the disposal of nuclear wastes in the sea, for the safe transport of radioactive materials, for the surveying and monitoring levels of radioactivity, and for safety regulations in ports and approaches used by nuclear merchant vessels. So far no regulatory actions have been taken, but on a regional level, the European Nuclear Energy Agency has supervised the dumping of radioactive wastes.

A similar process has been set in motion to cover the disposal of toxic chemicals. A Convention on the Control of Marine Pollution by Dumping from Ships and Aircrafts, banning the dumping of some substances and controlling others, has been prepared for adherence by nations bordering the northeastern Atlantic. A proposed international convention would forbid dumping of toxic wastes at sea without a governmental permit which would be issued only after a full statement of the nature of the wastes, their volume, the proposed methods of transportation, and the place where dumping is to take place. The information would, in addition, be passed on to a central international register so that the basis would exist for a continuous check on the scale of toxicity of the materials the oceans are being compelled to absorb. Such a register is, incidentally, still lacking for international dumping of oil from bilges and for the disposal of atomic wastes.

The limitations in these arrangements are clear enough. They all depend upon the voluntary cooperation of governments in imposing standards on their citizens. Better safety regulations and higher insurance—like superior navigational aids or better standards for the ship's crew—cost more money. There is thus a standing temptation to shipping companies to cut corners and costs and in this way gain a competitive advantage. In fact, the existence of "flags of convenience" by which shipowners register with Panama, Honduras, and Liberia precisely to avoid national regulations—the *Torrey Canyon* had Liberian registration—does not exactly promise total compliance with international standards. Indeed, more than ten years after the drawing up of the various conventions regulating oil emissions, as we have seen, at least two million tons of oil are still leaked to the sea each year.

For this reason, some governments are considering more seriously means of using both carrots and sticks—positive actions to get a better control of oceanic pollution, more effective deterrents against antiplanetary behavior. On the positive side there is, for instance, clearly scope for joint international action to move in, in a concerted way, on any large oil spill. Although no means are yet in sight for cleaning up the diffused global drift of oil clots from routine tank washings, methods and devices are being introduced by some nations to fight concentrated oil spills due to accidents. The Soviet Union has developed a special ship which is capable of skimming 7 tons an hour off the surface of the water. The Americans are working on an Air-Deliverable Transfer System complete with portable pumps and huge plastic bladders to contain the slicks. This can, apparently, go into operation within four hours of receiving a distress call and pump out 20,000 tons of oil within twenty-four hours of the signal. Such systems could be situated at strategic points along the oil routes and be ready on call. The suggestion has been made that their costs might be covered by a world insurance system. In this way, poor countries with coastlines as exposed as any other to the risk of massive pollution would not have to foot an impossible bill in cleaning up the mess.

This possibility is enhanced by another international development— the use of satellites as navigational guides. A ship in presatellite days would be lucky if it could get within a half mile of reporting its correct position. Now the pinpointing is so exact that men can return from the moon and splash down within a mile of the waiting helicopters. Accidental spills could be as exactly located and then speedily attacked. In fact, the question is beginning to be asked whether the international community would not be wise to transfer to the major trade routes the system now operating in the air—of traffic corridors into which craft come and then vary their pace only on permission from ground control. It has now been agreed that, in such an area of total congestion as the English Channel, ships must observe specific shipping lanes. The principle could be capable of useful extension. Moreover, satellite observers may before long be in a position to identify and photograph delinquent dumping of all kinds.

These activities have led to increasing interest in the possibility of some kind of international authority which could, on behalf of governments, oversee safety controls, set in motion disaster operations, and exercise a general policing function to see that ships do in fact obey the regulations laid down in the international conventions to which governments have agreed. Since there is no sovereignty over the oceans— outside the ominously growing "national waters"—an authority of the kind discussed could preserve the largest and least divisible part of man's planetary inheritance from the tribal squabbles and national divisions which have wasted, ruined, and soaked in blood so much of his territorial patrimony.

More comprehensive proposals have in fact been put forward to vest an international authority with supervisory rights over all waters beyond the depth of 200 meters and to include in those rights control over the seabed, supervision of the disarmament proposals already agreed for it, and perhaps disposal of whatever riches may be found there later. On this last point conflicts have arisen between states that wish to secure for themselves the minerals that may lie on contiguous continental shelves beyond the 200-meter line and the wider proposal that the potential wealth be held in trust for all the developing peoples. As methods of deep-sea mining extend the range of possible extraction, the division could become more acute. There is, therefore, all the more reason for reaching international agreement on jurisdiction before further advances to take over the last remaining areas in which the seamless web of air and water, of moving currents and great tides still lives and moves in its fundamental unity, still able to serve not this or that divisive interest by the common needs of all mankind.

The importance of this issue to the whole question of a workable planetary system can be underlined by reference to another potential area of responsibility for a possible maritime authority—monitoring the oceans from a series of regional research stations, with universally accepted standards and with total availability of the results to all governments, research bodies, and citizen groups. As we have seen, both with radioactive wastes and with dumping, we do not know the cumulative effect of what we are doing. It may seem inconceivable that a system as vast as the planet's interconnected seas, covering 70 per cent of the earth's surface and in constant interchange with the atmosphere and with the land masses about it, could ever be vitally or irreversibly affected by human activities. But an international body of marine biologists assembled at the 1971 Pacem in Maribus Conference concluded unanimously that marine life was in serious danger of destruction by pollution. Just as the balance in the atmosphere between planetary heating up and planetary cooling down is incredibly delicate, it may be that among the thermal exchanges of the great currents or the trace elements of critical minerals or the unimpeded living and breathing of minute phytoplankton, there are thresholds of stability which man could cross only at the risk of disaster. We are ignorant of all this. But as the weight we impose on both our airs and oceans steadily increases—the effluents, the transport, the recreation, the sheer pressure of multiplying populations—we have to know. And one of the principal ways in which a useful and reliable body of knowledge can be built up is by a very great increase in atmosphere and oceanic monitoring and by provision for international teams of experts to assess and interpret the facts on both global and regional bases.

The need for internationalizing such procedures was brought out clearly in the summer of 1971 when the Working Group on Marine Pollution met in London to prepare proposals for the Stockholm Conference on the Human Environment. At that meeting, eleven developing nations, drawn from Latin America, Africa, and Asia, called for a Third World monitoring system on the grounds that any research or survey of world-wide pollutants conducted by already developed powers could evolve into a further means of trade control and protectionism.

This is part of a wider concern. The first imperative for most developing countries is still for modernization and growth. This permanent need requires all the resources available for local use and developing countries are not likely to welcome the possibility of having to share in the cost of cleaning up oceanic pollutions they did not create. Moreover, development requires larger opportunities for international trade, not more restraints. Some developing governments fear that tighter environmental controls might be used as further barriers to their overseas sales.

How do we know, they argue, that this new concern with the environment is an honest one? We have seen our exports of foodstuffs excluded from developed markets on what has often seemed to us dubious, locally determined rules about sanitary standards. Will the risk of DDT now be used in the same way? Will standards of environmental control which we, as underindustrialized countries, do not yet require be made the precondition of admitting our exports? Will the development of local shipping be made even more impossibly expensive by the imposition of insurance costs and antipollution devices to clean up oceans which richer nations have corrupted for half a century without a thought for the environment? And if standards are not observed, are we as developing nations likely to find harbors closed to us and a new range of nontariff barriers put in the way of our exports, which represent less than 20 per cent of world trade and, in manufactures, less than 5 per cent?

The fact that the developed nations' increased interest in the human environment has coincided, among some of the wealthiest of them, with an apparent loss of concern for development assistance does little to allay such fears. No doubt there are new technologies which would permit the developing countries to by-pass the pollution stage and leapfrog into clean processing and clean power. No doubt the environmental risks inherent in building high dams and jumping into the middle of higher scientific monocultures can be offset by better environmental planning by growing cadres of technical expertise and wider scientific know-how. No doubt a decentralized plan for new towns and industries serving and stimulated by the Green Revolution offers a more desirable pattern than the likelihood of unrestricted, pell-mell urbanization in the next decades. No one denies that there can be better environmental answers. The difficulty is that many of them involve more capital, all of them require more skill, and none of them seems to be linked in any creative way with world strategies to permit developing countries to earn the needed capital or secure the extra skills. But at this point the issues go beyond the interdependence of nations in the planet's biosphere. The question turns on whether the technosphere—the constructed world order of technological innovation, investment flows, and commercial exchanges—can also be revised and managed to recognize the interdependence of nations and the underlying community of the species man.

14 COEXISTENCE IN THE TECHNOSPHERE

THE BLEAK MATHEMATICS OF population growth, of needed food supplies, of urban pressure, and of the capital sums required to mitigate their effects leave us in no doubt about the threat to humanity in developing lands. By the year 2000, their numbers will have grown by nearly three billion—or more than doubled in thirty years. To keep pace with this growth and to ensure a modest increase of income above the pitiful $100 a year of the great majority, governments must aim at economic growth of at least 6 per cent a year. To do this they must push up annual savings to well above 15 per cent a year of gross national product (the sum of goods and services), and do this in communities where, in many cases, 90 per cent of the people live little above margins of subsistence. Since much of the equipment, the technology, and the skills they need cannot yet be secured locally, parts of the needed investment must be in the shape of foreign exchange.

It is the considered view of every responsible international agency or commission that these targets are not attainable unless the developing nations have much greater access to developed markets and continue to receive a steady, and for a time increasing, flow of concessionary finance from the already wealthy lands.

The very respectable growth rates of the last decade were, as we have seen, to some extent secured by very large borrowing abroad. Today, the return flow of their payments to developed countries—of over $4 billion a year—is not much less than the concessionary finance made available by developed countries. This has stuck fast at about $6 billion and, in these times of increasingly inward-looking economic policies, shows little sign of increasing. Nor is the present mood of industrialized lands inclined to compensate for lagging capital flows by wider opportunities for trade.

The concessions to the developing world's manufactured exports negotiated by the United Nations Trade and Development Organization (UNCTAD) do not amount to more than a billion dollars a year. But the World Bank estimates a needed increase in exports from $7 billion to $28 billion a year by 1980 if growth targets are to be met. And all these calculations were made before any extra sums were added for essential environmental expenditure.

Were we to take seriously just three essential elements in a better environment—the backing up of the Green Revolution by popular participation and technical expertise, the linking to it of policies for decentralized labor-intensive industry and regional city building, and, perhaps most urgent of all, world resources sufficient to improve diets and, in particular, make good the protein deficiency among the poor countries' children, we could safely add another $100 billion a year in immediately necessary investment and at least double the expert skills needed to invest it wisely.

This is still only half the planet's lamentable annual budget of $200 billion for arms, which absorbs not only resources but trained manpower and scarce scientific expertise as well. In any case, a large share of the needed investment would be provided by developing nations themselves —so far 75 to 80 per cent of all investment for development has been raised locally. But the additional capital supplementing existing flows of aid and backed by training programs for development action would bring governments in developed lands to within sight of the 1 per cent of gross national product first suggested as the target for assistance in 1964 and half promised and half forgotten ever since. It would also begin to provide, both in developed and developing lands, the expert cadres needed for more rapid advance.

But the plain truth is that, on present evidence, neither the extra funds nor the extra openings for training nor the wider markets for trade are likely to be forthcoming. Indeed, a shrinking of existing opportunities is even possible. The degradation of human conditions in developing lands which must follow upon too low a rate of economic growth and social transformation is therefore likely to deepen in the seventies and to grow to catastrophic proportions in the 1980s.

As with the developed nations at the end of their first thirty years of rapid modernization—in the "hungry" 1840s—the obstructions seem to be outpacing the opportunities, the destruction of old ways appearing ahead of the creation of the new. It would be a bold prophet who, in such conditions, could look forward to anything but a deepening and spreading trend toward violence and anarchy in the wake of growing resentment and disappointed hope.

Yet during these same decades, there can be no doubt that the expectations of the developed peoples are likely to be centered on rising incomes and continued prosperity. If we carry on the curve of the 1950s and 1960s, we can easily forecast annual per capita incomes approaching $10,000 by the year 2000, and in the upper brackets, the two-home, three-car, four-TV-set family as the norm. Such standards could be the lot of a privileged elite of some one and a half billion developed peoples while for five billion others, an average income of $400 a year is the utmost reach of hope. And they, too, might be something of an elite while at the bottom of the social pyramid millions would be stunted by malnutrition and millions more scratching a bare living in filthy, workless cities and disintegrating countrysides. If developing peoples were as ignorant as Pharaoh's slaves of how "the other half" lives, they might toil on without protest. But the transistor and the satellites and world-wide TV have put an end to that kind of ignorance. Can we rationally suppose that they will accept a world "half slave, half free," half plunged in consumptive pleasures, half deprived of the bare decencies of life? Can we hope that the protest of the dispossessed will not erupt into local conflict and widening unrest?

Meanwhile, only a rough equilibrium of power, a precarious "balance of terror," underlies the relations of the world's great powers. It is only with difficulty that they can avoid being drawn into local conflicts and the more the conflicts erupt, the more their wisdom, diplomacy, and restraint may be taxed by the mounting pressures. Even today their fears are expressed in elaborate and still-proliferating weapon systems, in underground nuclear testing. Some powers, tragically, are still continuing to test in the air. These fears underlie the most dangerous of all pollutions, indiscriminate radiation, and could, if they escalate into a total confrontation, undermine all organized living and inflict grave, perhaps irreversible genetic damage on the few pitiful specimens who might survive the blast, the radiation, and the consequent visitations of world-wide famine and plague.

It is sometimes said of those who try to persuade man of his environmental predicament that they paint a picture so gloomy and irreversible that the average citizen's response is to go out and buy a can of beer. If nothing can be done to escape the onward rush of some irresistible eco-doom, then why take the trouble even to return the can? But indeed over a vast range of environmental problems, action *is* possible, policies *are* available, reversals *can* take place, water run clean, the sun shine over clear cities, the oceans cleanse our human shores, and harvests ripen in uncontaminated fields.

Indeed, some nations and other jurisdictions already are launched on effective planning and pollution-control programs. Already some cities enjoy cleaner air than they knew three or four decades ago. Rivers are being cleaned up and fish are returning to them. There are places where rangelands are managed properly, where soil erosion has been stopped, wildlife is preserved and timberland carefully reforested. There are even examples of reversing the deterioration of inner cities. And all this has been done within the limits of existing knowledge, known techniques, and institutional capabilities. There is, in fact, only one place where fear and doom are truly appropriate and that is when we confront man's oldest habit and most terrible institution—the organized, systematic killing of his own kind.

15 STRATEGIES FOR SURVIVAL

The Need for Knowledge

BUT WE ARE NOT sleepwalkers or sheep. If men have not hitherto realized the extent of their planetary interdependence, it was in part at least because, in clear, precise physical and scientific fact, it did not yet exist. The new insights of our fundamental condition can also become the insights of our survival. We may be learning just in time.

There are three clear fields in which we can already begin to perceive the direction in which our planetary policies have to go. They match the three separate, powerful and divisive thrusts—of science, of markets, of nations—which have brought us, with such tremendous force, to our present predicament. And they point in the opposite direction—to a deeper and more widely shared knowledge of our environmental unity, to a new sense of partnership and sharing in our sovereign economics and politics, to a wider loyalty which transcends the traditional limited allegiance of tribes and peoples. There are already pointers to these necessities. We have now to make them the new drives and imperatives of our planetary existence.

We can begin with knowledge.

The first step toward devising a strategy for planet Earth is for the nations to accept a *collective* responsibility for discovering more—much more—about the natural system and how it is affected by man's activities and vice versa. This implies cooperative monitoring, research, and study on an unprecedented scale. It implies an intensive world-wide network for the systematic exchange of knowledge and experience. It implies a quite new readiness to take research wherever it is needed, with the backing of international financing. It means the fullest cooperation in converting knowledge into action—whether it be placing research satellites in orbit or reaching agreements on fishing, or introducing a new control for snail-borne disease.

But it is important not to make so much of our state of ignorance that we are inhibited from vigorous action now. For while there is much that we do not yet understand, there are fundamental things that we *do* know. Above all, we know that there are limits to the burdens that the natural system and its components can bear, limits to the levels of toxic substances the human body can tolerate, limits to the amount of manipulation that man can exert upon natural balances without causing a breakdown in the system, limits to the psychic shock that men and societies can absorb from relentlessly accelerating social change—or social degradation. In many cases we cannot yet define these limits. But wherever the danger signals are appearing—inland seas losing oxygen, pesticides producing resistant strains of pests, laterite replacing tropical forests, carbon dioxide in the air, poisons in the ocean, the ills of the inner cities —we must be ready to set in motion the cooperative international efforts of directed research which make available, with all possible speed, solutions for those most intimately concerned with the immediate problem and wider knowledge for all men of how our natural systems actually work. To go blindly on, sharing, inadvertently, the risks and keeping to ourselves the knowledge needed for solutions can only mean more agonies than we can cope with and more danger than future generations deserve.

A full and open sharing of new knowledge about the interdependence of the planetary systems on which we all depend can also help us, as it were, to creep up on the infinitely sensitive issues of divisive economic and political sovereignty.

Sovereignty and Decision-Making

Given our millennial habits of separate decision-making and the recent tremendous explosion of *national* power, how can any perception of the biosphere's essential unity and interdependence be combined with the acutely self-conscious separate sovereignty of more than 130 national governments?

Yet, in fact, for at least a century, some habits of cooperation have been accepted by states simply through recognition of their own self-interests. Ever since the world economy began to increase in extent and interdependence in the eighteenth and nineteenth centuries, sovereign states have shared some of their authority either by binding themselves to certain forms of cooperative behavior or by delegating limited power to other bodies. Despite rhetorical insistence on absolute sovereignty, governments have recognized in practice that this is impossible in some cases and inordinately foolish in many more. It is no use claiming the sovereign right not to deliver other people's letters if they use their sovereign right to refuse yours. The alternative to international allocation of radio frequencies would be chaos in world communications to the disadvantage and danger of all states. In brief, when governments are faced with such realities, they have exercised their inherent sovereign right to share voluntarily their sovereignty with others in limited and agreed areas of activity.

In the twentieth century, as a consequence of an ever greater overlap between supposedly sovereign national interests, the number of international treaties, conventions, organizations, consultative forums, and cooperative programs has multiplied rapidly. The growth of an intergovernmental community finds its most concrete expression in the United Nations and its family of specialized functional agencies and regional commissions. Outside the United Nations system, there has been an analagous growth of international organizations, governmental and nongovernmental, especially on the regional level.

All intergovernmental institutions are still, ultimately, creatures of national governments, but a large amount of their day-to-day work is sufficiently and obviously useful that a measure of authority and initiative comes to rest with them. They acquire support within national governments from the relevant ministries and agencies which, in turn, find useful constituencies within the ranks of international organizations. This is, none of it, a formal departure from sovereignty. But a strict, literal definition of sovereignty gets blurred in practice and the existence of continuous forums for debate and bargaining helps instill the habit of cooperation into the affairs of reluctant governments.

It is on to this scene of ultimate national sovereignty and proliferating intermediate institutions that the new environmental imperatives have broken in the last few years. The first effect has undoubtedly been to complicate still further a very complicated situation. Quite suddenly, for a whole variety of reasons, a very wide range of institutions have added an environmental concern to their other interests. In some cases, traditional programs and activities have been renamed to qualify them under the environmental rubric. In others a number of agencies have taken up the same environmental topic, though mainly from differing points of view. There has been some genuine innovation, and there is much ferment and groping as international organizations, to a greater or lesser degree, seek to comprehend and to adapt to the environmental imperative.

One example of combined good will and overlap can be taken from air pollution. The industrialized nations are the main polluters. So regional groupings are starting to respond. The Organization for Economic Cooperation and Development—the successor to the old Marshall Plan bureaucracy, linking North America with Western Europe and, more recently, Japan—is setting up an Environment Committee to coordinate a number of its existing research activities, for instance, its Air Management Research Group. The regional commissions of the United Nations are also beginning to move and the Economic Commission for Europe also has a Committee of Experts on Air Pollution. So has the North Atlantic Treaty Organization, which includes air pollution among a number of other research activities such as open waters and inland waters pollution, disaster relief, and regional decision-making for environmental issues.

This picture of somewhat uncoordinated and hence not fully focused activity, however, largely reflects the recentness of the environmental awareness. National governments, too, are trying to find means of adding an environmental angle of vision to institutions which have hitherto followed the traditional one-track approach to specialized problems through separate and usually uncoordinated administration. A rash of environmental councils and commissions is now appearing round the world to coordinate the activities of hitherto separate ministries. Several countries have taken the bolder step of bringing relevant ministries—housing, transport, technology—together in single Departments of the Environment. The various experiments are mostly not yet two years old, and it is too soon to say how well they may succeed in introducing an integrative view of man-environment relations into the national decision-making processes. Certainly it will not be easy.

And certainly it will be still more difficult at the international than at national levels of decision-making. So locked are we within our tribal units, so possessive over national rights, so suspicious of any extension of international authority that we may fail to sense the need for dedicated and committed action over the whole field of planetary necessities. Nonetheless there are jobs to be done which perhaps require at this stage no more than a limited, special, and basically self-interested application of the global point of view. For instance, it is only by forthright cooperation and action at the global level that nations can protect mankind from inadvertent and potentially disastrous modification in the planetary weather system, over which no nation can assert sovereignty. Again, no sovereignty can hold sway over the single, interconnected global ocean system which is nature's ultimate sink and man's favorite sewer.

Where pretentions to national sovereignty have no relevance to perceived problems, nations have no choice but to follow the course of common policy and coordinated action. In three vital, related areas this is now the undeniable case—the global atmosphere, the global oceans, and the global weather system. All require the adoption of a planetary approach by the leaders of nations, no matter how parochial their point of view toward matters that lie within national jurisdiction. A strategy for planet Earth, undergirded by a sense of collective responsibility to discover more about man-environment relations, could well move, then, into operation on these three fronts: atmosphere, oceans, and climate. It is no small undertaking, but quite possibly the very minimum required in defense of the future of the human race.

But it is not only the pollutions and degradations of the atmosphere and the oceans that threaten the quality of life at the planetary level. There are threats, too, of disease spreading among undernourished children, of protein deficiency maiming the intelligence of millions, of spreading illiteracy combined with rising numbers of unemployed intellectuals, of landless workers streaming to the squalid cities, and worklessness growing there to engulf a quarter of the working force. An acceptable strategy for planet Earth must, then, explicitly take account of the fact that the natural resource most threatened with pollution, most exposed to degradation, most liable to irreversible damage is not this or that species, not this or that plant or biome or habitat, not even the free airs or the great oceans. It is man himself.

The Survival of Man

Here again, no one nation, not even groups of nations, can, acting separately, avoid the tragedy of increasing divisions between wealthy north and poverty-stricken south in our planet. No nations, on their own, can offset the risk of deepening disorder. No nations, acting singly or only with their own kind, rich or poor, can stave off the risk of unacceptable paternalism on the one hand or resentful rejection on the other. International policies are, in fact, within sight of the point reached by *internal* development in the mid-nineteenth century. Either they will move on to a community based upon a more systematic sharing of wealth —through progressive income tax, through general policies for education, shelter, health, and housing—or they will break down in revolt and anarchy. Many of today's proposals for development aid, through international channels, are a first sketch of such a system.

But at this point, if gloom is the psychological risk of all too many ecological forecasts, may we not go to the opposite extreme of Pollyanna optimism in forecasting any such growth of a sense of community in our troubled and divided planet? With war as mankind's oldest habit and divided sovereignty as his most treasured inheritance, where are the energies, the psychic force, the profound commitment needed for a wider loyalty?

Loyalty may, however, be the key. It is the view of many modern psychologists that man is a killer not because of any biological imperative but because of his capacity for misplaced loyalty. He will do in the name of a wider allegiance what he would shrink to do in his own nature. His massive, organized killings—the kind that distinguishes him from all other animals—are invariably done in the name of faith or ideology, of people or clan. Here, it is not wholly irrational to hope that the full realization of planetary interdependence—in biosphere and technosphere alike—may begin to affect man in the depths of his capacity for psychic commitment. All loyalty is based on two elements—the hope of protection and the hope of enhancement. On either count, the new ecological imperative can give a new vision of where man belongs in his final security and his final sense of dignity and identity.

At the most down-to-earth level of self-interest, it is the realization of the planet's totally continuous and interdependent systems of air, land,

and water that helps to keep a check on the ultimate lunacies of nuclear weaponry. When after the nuclear testing conducted in 1969, the air above Britain was found to contain 20 per cent more strontium 90 and cesium 137, it is not a very sophisticated guess that the air of the testing states contained no less. It is the force of such recognitions that lay behind the first global environmental agreement—the Test-Ban Treaty negotiated in 1963—which has kept earlier nuclear powers out of competitive air testing and saved unnumbered children from leukemia. Similar calculations of enlightened self-interest underlie the treaty to keep nuclear weapons out of space, off the seabeds, and away from Antarctica.

Where negotiations continue—as in the Treaty to prevent the spread of nuclear weapons, or the Soviet-American talks on a mutual limitation of strategic arms—the underlying rationale is still the same. As the airs and oceans flow round our little planet, there is not much difference between your strontium 90 and my strontium 90. They are lethal to us both.

It is even possible that recognition of our environmental interdependence can do more than save us, negatively, from the final folly of war. It could, positively, give us that sense of community, of belonging and living together, without which no human society can be built up, survive, and prosper. Our links of blood and history, our sense of shared culture and achievement, our traditions, our faiths are all precious and enrich the world with the variety of scale and function required for every vital ecosystem. But we have lacked a wider rationale of unity. Our prophets have sought it. Our poets have dreamed of it. But it is only in our own day that astronomers, physicists, geologists, chemists, biologists, anthropologists, ethnologists, and archaeologists have all combined in a single witness of advanced science to tell us that, in every alphabet of our being, we do indeed belong to a single system, powered by a single energy, manifesting a fundamental unity under all its variations, depending for its survival on the balance and health of the total system.

If this vision of unity—which is not a vision only but a hard and inescapable scientific fact—can become part of the common insight of all the inhabitants of planet Earth, then we may find that, beyond all our inevitable pluralisms, we can achieve just enough unity of purpose to build a human world.

In such a world, the practices and institutions with which we are familiar inside our domestic societies would become, suitably modified, the basis of planetary order. In fact, in many of our present international institutions the sketch of such a system already exists. A part of the process would be the nonviolent settlement of disputes with legal, arbitral, and policing prodecures on an international basis. Part of it would be the transfer of resources from rich to poor through progressive world sharing—the system of which a 1 per cent standard of gross national product for aid-giving is the first faint sign. World plans for health and education, world investment in progressive farming, a world strategy for better cities, world action for pollution control and an enhanced environment would simply be seen as logical extensions of the practice of limited intergovernmental cooperation, already imposed by mutual functional needs and interests.

Our new knowledge of our planetary interdependence demands that the functions are now seen to be world-wide and supported with as rational a concept of self-interest. Governments have already paid lip service to such a view of the world by setting up a whole variety of United Nations agencies whose duty it is to elaborate world-wide strategies. But the idea of authority and energy and resources to support their policies seems strange, visionary, and Utopian at present, simply because world institutions are not backed by any sense of planetary community and commitment. Indeed, the whole idea of operating effectively at the world level still seems in some way peculiar and unlikely. The planet is not yet a center of rational loyalty for all mankind.

But possibly it is precisely this shift of loyalty that a profound and deepening sense of our shared and interdependent biosphere can stir to life in us. That men can experience such transformations is not in doubt. From family to clan, from clan to nation, from nation to federation— such enlargements of allegiance have occurred without wiping out the earlier loves. Today, in human society, we can perhaps hope to survive in all our prized diversity provided we can achieve an ultimate loyalty to our single, beautiful, and vulnerable planet Earth.

Alone in space, alone in its life-supporting systems, powered by inconceivable energies, mediating them to us through the most delicate adjustments, wayward, unlikely, unpredictable, but nourishing, enlivening, and enriching in the largest degree—is this not a precious home for all of us earthlings? Is it not worth our love? Does it not deserve all the inventiveness and courage and generosity of which we are capable to preserve it from degradation and destruction and, by doing so, to secure our own survival?

INDEX